Promoting Property

CU00519581

This book explores the wide-ranging elements of property PR in the UK, with a strong emphasis on communications theory, strategy and technique. The editors begin with an introduction to the property cycle and the role of property PR within it; consideration of the changes and challenges facing the industry; various structures of property communications; and the need for a strategic approach.

Subsequent chapters provide perspectives and lessons from contributors in a variety of property sectors including commercial property, estate agency, social housing, property consultancy, proptech, retail and home-building. The book concludes with insight into future change, both for the property industry and for the communication function within it.

This book is recommended reading for all property PR teams, for students studying for property, PR or marketing degrees, and for anyone working in the built environment sector who needs to consider PR and marketing as part of their role.

Penny Norton is the director of PNPR and founder of *ConsultOnline*. Her work covers all elements of property PR, from pre-planning consultation through to media relations for property consultancies. Penny is the author of *Public Consultation and Community Involvement in Planning: A Twenty-First Century Guide* (Routledge, 2017) and she writes extensively for property publications.

Liz Male MBE is the founder of LMC (*Liz Male Consulting*). She has 30 years' experience of marketing communications, corporate reputation and issues-led PR for the property and construction industry in both agency and in-house roles.

'Property is a fast-paced sector with multiple and complex communications needs. As such it is dependent upon good PR, and also offers wide-ranging opportunities to CIPR members. A "bible" of property PR is long overdue and it's great to see so much expertise featured in this book. Providing such a comprehensive overview of the sector, together with a strong emphasis on theory, strategy and techniques, is a considerable achievement and will be of great value to PR practitioners.'

Emma Leech, CIPR President

'A fascinating and thorough insight into the world of property promotion through the eyes of those who know it best. A must-have guide.'

Laura Stevens, Interim Head of Marketing, Lambert Smith Hampton

'This collection of essays provides some real nuggets of insight and analysis. So much of what is written here needs to be listened to and acted upon.'

Liz Peace, previously CEO of the British Property Federation, now an independent director, chair, and advisor to the industry

'Working on large-scale developments is incredibly complex and carries huge responsibility, so investing in strategic communications is vital to tell a positive story about the creation of a new place, manage multiple stake-holders with sensitivity and often deal with unpredictable events. As we grasp with climate change, technological disruption and changing demographics, I hope this much-needed book attracts more PR experts to the exciting, vital practice of property communications.'

Shaun Harley, Director of Communications, Homes England

'This book provides valuable insight into the various challenges currently facing the built environment industry and how the public relations industry is seeking to address them. Property and construction, in particular, have some significant issues to confront, both in terms of perception and reality, and *Promoting Property* doesn't shy away from confronting them.'

Adam Branson, property journalist

Promoting Property

Insight, Experience and Best Practice

Edited by Penny Norton and Liz Male

Routledge
Taylor & Francis Group

LONDON AND NEW YORK

First published 2020
by Routledge
2 Park Square, Milton Park, Abingdon, Oxon OX14 4RN

and by Routledge
52 Vanderbilt Avenue, New York, NY 10017

Routledge is an imprint of the Taylor & Francis Group, an informa business

© 2020 selection and editorial matter, Penny Norton and Liz Male;
individual chapters, the contributors

British Library Cataloguing-in-Publication Data
A catalogue record for this book is available from the British Library

Library of Congress Cataloging-in-Publication Data
A catalog record has been requested for this book

ISBN: 978-0-367-25716-3 (hbk)
ISBN: 978-0-367-25717-0 (pbk)
ISBN: 978-0-429-28936-1 (ebk)

Typeset in Goudy
by Wearset Ltd, Boldon, Tyne and Wear

Contents

Contributors

Penny Norton is director of PNPR and has worked with leading property companies, commercial developers, both private and social housing providers, architects, interior designers and local authorities. Clients have included CBRE, Carter Jonas, British Land, Chestertons and Broadgate Estates, More London and Sainsbury's. Penny has written extensively on the subject of communications for property publications and her book, *Public Consultation and Community Involvement in Planning: A Twenty-First Century Guide* was published by Routledge in 2017. Penny is a Fellow of the Chartered Institute of Public Relations and founded the CIPR's Construction and Property Special Interest Group.

Liz Male is founding director of LMC (Liz Male Consulting), a PR and communications consultancy specialising in construction, property and the built environment. She has more than 30 years' experience of advising private and public sector organisations about how best to enhance their reputations, and also has wide-ranging experience as a non-executive director. Liz was chair of TrustMark until 2017 and was on the Construction Leadership Council and the implementation board of the Government's Each Home Counts review which developed new standards for the low carbon retrofit of existing homes. She currently chairs the National Energy Foundation and is a lay board member of the Architects Registration Board (ARB). Liz was awarded an MBE in the 2015 New Year's Honours for services to construction and consumer protection.

Emma Drake is an independent communications consultant and Chartered Public Relations Practitioner who started her career in corporate communication and has since gained 20 years' experience in the PR and communications industry, more than half of which has been within the built environment sector. Emma is also founder of Henbe, which provides corporate communication, marketing, and PR consultancy to organisations in the built environment, predominantly in housing and large-scale development, but also finance and technology-led products and professional services firms servicing this sector. Emma has been an advocate of best practice techniques and professionalising the PR industry her entire career and was one of the first Chartered Institute of Public Relations (CIPR) members to obtain the Postgraduate Diploma from the CIPR and, subsequently, the Crisis Communications Diploma. She is a volunteer director of a local community charity, a mentor to those entering the industry, and also a member of the Chartered Institute of Marketing.

Dan Innes is the founder of Innesco. He has worked in real estate and the built environment for over 25 years, spent entirely in an agency environment aside from the first year after graduating, which he spent working in the real estate planning team at what is now AXA. He works to articulate success, build reputations, trigger business growth and drive asset value. He and the wider Innesco business have gained the trust of numerous property clients, including AXA, Blackrock, Brookfield, Capital & Regional, CBREi, AlliedLondon, Hammerson, Hermes, HINES, Landsec, Bouygues, Pradera, Delancey, IKEA Centres, Stanhope, intu Spain, British Land, MAPIC, Westfield, Tesco, The Crown Estate, and The Wellcome Trust.

Clare Jones is founding director of Grasshopper Communications. She has over a decade's experience in delivering communications campaigns for property and regeneration projects, including a number of mixed-use urban extensions and homebuilding projects. She has particular expertise in designing and implementing community and stakeholder engagement programmes for major planning and infrastructure projects, and is interested in best practice approaches to grassroots community engagement. She is a member of the CIPR and is currently vice chair of the South Wales branch of Women in Property.

Susan Fox is a communications consultant and coach. She was director of communications for The Regenda Group, where she led a major project to rebrand the Group and its subsidiaries. Previously, as director of policy and communications, Susan established the policy function at Ombudsman Services, and as director of corporate affairs at the Information Commissioner's Office, she achieved the highest ever awareness of data protection and freedom of information. She is a Chartered PR Practitioner and a Fellow of the CIPR, and contributed to the CIPR book *Chartered Public Relations*.

Fiona Lund's 25-year career has been defined by strategic communications and the delivery of professional public relations and marketing across the housing supply chain. For the last 13 years she has led Brouha, an award-winning agency, to becoming one of the leading strategic boutique PR brands in social housing, retail, housebuilding and property services.

Henry Columbine is a managing partner of Newgate Communications. He has 12 years' experience in strategic communications, with a focus on the property sector. His work includes corporate campaigns for real estate developers, investors and consultants, as well as supporting sales and lettings programmes for commercial, residential and mixed-use schemes.

Tania Thomas and Henrietta Harwood-Smith are joint founders of Maison Communications, a London-based boutique agency that specialises in property, lifestyle, interiors and architecture, with a passion for design and wellbeing. Maison provides corporate and consumer campaigns for start-ups through to global brands and listed PLCs. Clients have included Great Portland Estates, The Buying Solution (the buying arm of Knight Frank), The Modern House, Ekkist and Banda Property. Tania was previously head of the property division at Luchford and prior to that Head of Residential PR at Knight Frank. Henrietta worked at Luchford before heading up Hamptons International's PR.

Tricia Topping has over two decades' property marketing experience across the residential and commercial sectors. Previously chief executive of TTA Group (now Chime Communications subsidiary Good Relations Property), she founded Carlyle Consultants in 2013. Her love of property, art and interior design inspired Tricia to create Luxury Topping, a newsletter which focuses on 'the business of luxury'. She is a Fellow of the Royal Society of Arts, a member of the Directors' Circle of the Victoria & Albert Museum, a Freeman of the City of London, and judge of both the International Property Awards and the What House? Interior Design Awards. She won Marketing Firm of the Year at the Influential Businesswoman Awards in 2018 and Best Luxury Product Marketing Agency, South East in 2019.

Jamie Jago began her PR career in 1987 as an in-house PR manager at a small boutique estate agency in London. Three years later she set up Jago D'Arcy Hale, specialising in residential property in the UK and abroad. In 2002 Alison Dean joined as co-owner and the firm was rebranded Jago Dean. Widely regarded as the leading PR firm in the industry, it was acquired in 2008 by Savills, where she remains to this day, running the B2C press office.

Ed Mead has been hands-on in the property sector since 1979. He has been director of various companies, most recently Douglas and Gordon. He has written weekly columns as Agent Provocateur in *The Daily Telegraph* and as the *Sunday Times* Property Expert. He has fronted two BBC TV property series, has sat on the board of The Property Ombudsman and is a Fellow of the RICS. In 2016 he launched Viewber, which has given him a deep insight into the rapidly expanding world of proptech.

Louise Parr is a senior PR and marketing manager who specialises in technology marketing – specifically product launches and creating new markets for large global companies and growing tech firms. Her background includes global campaign development and product launches and execution for global leaders Cisco Systems, Vodafone, EE and BlackBerry.

Foreword

A strategic approach to property PR

We have strange and contradictory feelings about buildings. There is almost universal agreement that the country needs more new houses, yet property developers are almost invariably cast as villains in popular narrative. New buildings, like the Gherkin and the Shard, are fiercely resisted until they go up, yet once completed they quickly become well-loved friends. Our New Towns are now old, but long-derelict old buildings can power enormous regeneration schemes such as Bankside and Battersea.

In this strange world, quite literally everyone has a stake – the Government, landowners, large commercial building firms, foreign investors, banks, pension funds, environmentalists, local businesses and individual home-owners. Everyone has strong feelings about the place where they live, and proposals that will have an impact on those places affect those feelings deeply. This is the fundamental reason why property needs public relations. While the technical and financial aspects of the property sector may lie outside most people's interests, everyone wants to have a say about the places that matter enormously to them.

This book serves as an excellent primer to those planning a career in public relations, showing the extraordinary scope that exists within roles in the property sector. If you are unsure where you want to build your career, there is much to consider in property public relations, and this book will give you a solid grounding in today's key issues and challenges.

Equally, for those working in other parts of the property sector, this book is a handy reference for how public relations professionals can help you succeed. Increasingly, those playing other roles in the property sector are coming to appreciate the power of their publics and the importance of good relations with them. Indeed, there can be few areas in which the importance of public relations to continued success is so critical. Property is a sector where organisations carry substantial financial and reputational risks, and strategic public relations has an essential role to play in the long-term sustainability of the business.

Effective organisations and their leaders are purpose driven. They must ensure that their values, strategy and culture align with that purpose. They also need to foster effective stakeholder relationships aligned with that purpose. Dialogue, a critical part of stakeholder management, is one of the underlying principles of developed public relations practice. Decision-making is improved when an organisation can understand and respond to the concerns of important groups – including its own workforce, its supply chain, and indeed, those whose interests may bring them into direct conflict. Engaging with government, planning authorities, amenity groups and others is essential for success. This approach provides feedback to challenge corporate behaviour as well as testing and improving how decisions are made.

For these reasons, it is inconceivable that a successful organisation would try to develop and deliver a strategy without public relations. There is no organisational objective that is not dependent to some extent on good communication and relationships. Public relations builds value for organisations through its contributions to intellectual, human, reputational and relationship capital, and through its contribution to the management of risk and uncertainty, critical decision-making and effective delivery of plans.

To help organisations bring in the best public relations advice, the CIPR publishes a recruitment guide with advice on how to select the right candidate for a role. It also publishes a client guide for organisations seeking agency support rather than recruiting for in-house positions. The CIPR's own Construction and Property Special Interest Group is a forum where those working in the sector can develop their knowledge and their personal network. If you enjoy reading this book, I hope you will come along.

Alistair McCapra
Chief Executive of the Chartered Institute of Public Relations

Preface

'Property PR? That sounds very niche' was a comment that we received more than once when we announced that we were editing this book.

But this could not be more wrong. Property is an extensive and incredibly varied industry, as is the PR and communications expertise that serves it. Even within the scope of this book we have merely scratched the surface of property PR and all its technicolour glory.

That said, the colour, depth and diversity create a fascinating glimpse into the practice of property PR and, we hope, provide a comprehensive introduction as well as some expert knowledge and best practice. We are very grateful to the contributors who came forward with their insights, case studies and personal stories from a variety of property sectors.

The book is invariably slightly biased towards residential property: looking at new homes in the private and public sectors and the marketing and sale of existing homes at all points in the price range. We also have a chapter on commercial and retail property from a leading PR agency in the sector, and some very interesting chapters relating to the use of technology in property, PR for interior design and the promotion of property consultancies. There are, of course, property functions that are not covered in this book – but we hope to include them in a second edition.

Arguably, property PR is not as pioneering as PR for some other sectors and is sometimes fairly traditional. But it is increasingly becoming a strategic management discipline, and consequently, PR practitioners are getting a stronger voice at board level.

This is exactly as it should be. Property – the communities we create and the buildings in which we live, work, rest and play – is by far the biggest factor in our health and wellbeing, safety and security, environmental responsibility and social cohesion. To be an influential PR and communications professional in this sector is to make a genuine difference to people's lives.

As property PR continues to develop in strategic sophistication, creativity and ever-extensive methodology, it will be fascinating to see how it evolves.

In the meantime, we hope that this book provides an interesting and educational resource for all property PR teams, students studying for property, PR or marketing degrees, and for anyone working in the built environment sector who needs to consider PR and marketing as part of their role.

We see this book (along with its companion *Communicating Construction: Insight, Experience and Best Practice*) as part of a continual learning experience, and we hope to continue the discussions on LinkedIn, where we would welcome comments, thoughts and general feedback.

Penny Norton and Liz Male
October 2019

Acknowledgements

The authors would like to thank the following for sharing their knowledge and insight: Ian Anderson (Cushman & Wakefield), Alison Blease (Hamptons International), Rebecca Britton (Urban and Civic), Richard Brown (The Regenda Group), Victoria Buchanan (Savills), Michelle Carvill (Carvill Creative), Asif Choudry (Resource), Matthew Clark (Key Property Marketing), Andrew Clark (David Phillips), Amanda Coleman (Greater Manchester Police), Richard Collingwood (The Riverside Group), David Cox (ARLA Propertymark), Paul Dobbie (Persuasion PR), Daniel Flood (HMLR), Grace French (Stand Agency), Alexei Ghavami (Inspired Homes), Mike Goulding (Homes England), Andy Green (Story Starts Here), Michelle Harris (Harris & Home), Bobbie Hough (Hough Bellis Communications), Tom Hustler (Homes England), Neil Johnson (HMLR), Katie Jones (The Regenda Group), Christine Lalumia (Victoria & Albert Museum, Sotheby's), Jo Leckie (Good Shepherd), Ben Lee (Bidwells), Paula Lent (Artmasters), Ben Lowndes (Social Communications), Matthew Marshall (Redwing Living), Becky Merchant (Stand Agency), Chris Morgan (Kamma), Paul Mumford (Places for People), John Neugebauer, Laura Oliphant (Stand Agency), Gwyn Owen (Essex Housing), Clare Paredes (National Housing Federation), Mary Parsons (Places for People), David Patterson (Innesco), Nina Peters (The Regenda Group), Jeremy Porteus (Housing LIN), James Rae (The Riverside Group), Peter Sawa (Westgreen Construction), Danielle Sharp (Plus Dane Housing), Orla Shields (Kamma), Kieron Smith (FTI), Laura Stevens (Lambert Smith Hampton), Simon Stretch (Innesco), Andrew Taylor (Countryside Properties), Katie Teasdale (National Housing Federation), Heather Topel (North West Cambridge Development), Andrew Trigg (HMLR), Mike Turner (Ian Williams) and Nick Whitten (JLL).

1   Introduction

Penny Norton and Liz Male

Property: an extensive and varied industry

The world's real estate is worth US$280.6 trillion[1]: on average 3.5 times gross domestic product (GDP) and significantly more so in Europe, China/ Hong Kong and North America. In the UK, the total stock of property assets has grown at an average rate of 6.1 per cent over 30 years[2] – close to double the rate of inflation – primarily due to a significant rise in the value of residential property.

Unsurprisingly, then, over those 30 years a lucrative market for public relations (PR) in property has continued to grow. Of the UK PR and communications industry – itself worth £14.9 billion and employing more than 95,000 people[3] – approximately 17 per cent now work in property and construction.[4]

Given the size and diversity of the property industry, property PR is inevitably wide-ranging, encompassing most of the industry's skills and specialisms. The many different sub-disciplines of PR – which include business to business PR, community relations, consumer PR, corporate PR, crisis management, corporate social responsibility (CSR), education and arts work, sponsorship, financial PR, local government lobbying, media relations, public consultation, social and online media, and stakeholder engagement – are deployed at most stages.

This book, and its partner book *Communicating Construction: Insight, Experience and Best Practice*,[5] explains and showcases how these PR disciplines operate at each stage of the property cycle.

The property process consists of various stages, as shown in Figure 1.1, but with each intrinsically connected to the next. There is no isolated discipline or sector, or indeed any single property sector. For this reason, it is vital that a communications professional working in a specific field has a good understanding of the broader context.

Furthermore, the end product varies considerably. This book covers the function of promoting a large-scale mixed-use scheme and the variety of property types that may exist within such a scheme (commercial and retail, private housing, social housing, student accommodation, build to rent, and

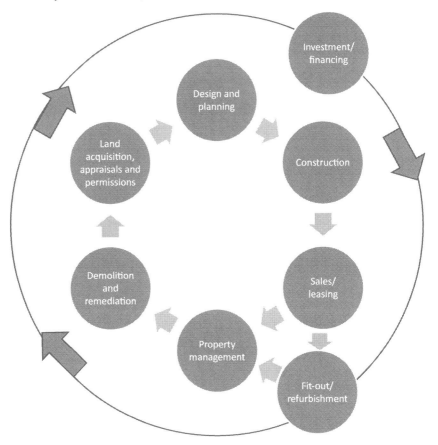

Figure 1.1 The property lifecycle.[6]

the 'top end' of the residential market). It goes on to cover interior design and looks at how properties are marketed, with sections on estate agency (both on and offline) and the growing function of proptech. A chapter on property consultancy describes the communications function of companies that provide advice and services at each stage of the property process.

Communications professionals working within any of these disciplines are advised not only to read about their own area but to gain a greater understanding of the PR practices in other specialisms.

While sectors as diverse as social housing and interior design, master-planning and proptech might appear worlds apart, the essential communica-tions skills – the basis of research and understanding; planned, two-way, clear and transparent communication; monitored and evaluated – remain constant, and there is no better way of benefiting understanding of one sector than through a knowledge of the bigger picture.

Change and challenge

External influences

Property is an exciting sector because of the way in which it is impacted by (and influences) social, technological, political and economic change.

Nothing stands still in property. As Table 1.1 shows, property PR is impacted by an extremely wide range of external influences and must consistently adapt and respond.

Modernisation from within

Simultaneously, the property industry is under considerable internal pressure to modernise itself, and the communications function is instrumental in delivering and promoting change.

A report carried out by the Quoted Companies Alliance and BDO LLP[7] shows that a significant proportion of a company's value – around 28 per cent – is accounted for by its reputation. And in an industry where stakeholder voices are increasingly heard – from residents only too happy to leverage various online channels (ApartmentRatings.com, Google Reviews, Facebook and others) to communicate their displeasure and frustration with property issues, to the stock market, which immediately reflects damaged reputation – it is vital that public profiles are safeguarded.

The housebuilding industry in particular has suffered from reputational damage in relation to poor standards,[8] executive pay[9] and leasehold abuses.[10] More generally, the property industry has brought reputational damage upon itself through some of its outdated and morally questionable activities at both its annual global conference, MIPIM[11] and at an exclusive charitable function, The President's Club.[12]

Environmental responsibility also poses a reputational challenge. Buildings account for 32 per cent of global energy use and 19 per cent of greenhouse gas emissions.[13] The commitment to improve environmental sustainability in UK property and construction has been significant, yet greater effort is needed to meet the urgency of the climate change situation and society's expectations. And again, the challenge is not only in bringing about the change: it is in communicating positive change and in doing so, educating, challenging preconceptions and supporting broader environmental progress.

Another issue for the property sector is recruitment and retention, and again, communications is vital in addressing this. Property development and its construction supply chain have been tangibly impacted by a shortage of suitable workers. Despite some very positive steps on behalf of the Royal Institution of Chartered Surveyors (RICS) and industry groups (Freehold, Planning Out, Women in Property, Women in Construction and others), the public perception of the property sector is that of a

Table 1.1 Sample external factors influencing property PR

Social	• Rising consumer demand for housing, associated facilities and infrastructure • Demographic change: the rise of Generation X (the 'experience generation') and the 'silver tsunami' resulting in changing expectations of housing and retail/leisure facilities • Social change leading to the introduction of new property assets – including co-working, pop-ups and co-living, retirement villages, and an increase in student housing as a result of more young people attending university • Changing opinion – for example, on climate change • Changing consumer demands – affecting the layout and composition of residential units • A requirement for a greater emphasis on social responsibility and wellbeing in the built environment – such as WELL* Version 2**
Political	• General elections, local elections and pre-election purdah • Emerging policy – on infrastructure, industry, housing, planning and regional economic programmes such as the Northern Powerhouse and the Oxford to Cambridge Arc • Impact of policy reviews (Task Forces, White Papers) on sentiment • Political pressure – such as the requirement to meet housing targets • The wide-ranging impact of Brexit – on issues including recruitment and retention of skills, sourcing of materials, European legislation and property values • Changing structures for funding affordable housing
Regulatory	• Changes in legislation – for example, the new fire safety standards that followed the Grenfell Tower fire and Dame Judith Hackitt's Independent Review of Building Regulations and Fire Safety • Reviews and updates to Building Regulations and other UK, EU and international building standards
Financial	• The financial cycle – growth and recession • The financial year – annual results and financial reports • Lack of resources in sectors including planning and affordable housing
Environmental	• Changing requirements for sustainable initiatives; response to climate change (e.g. flood defences) • Changing environmental regulations – including Minimum Energy Efficiency Standards, Energy Performance Certificates and Part L of the Building Regulations. Expectations of a move towards a circular economy in which recycling features prominently • Environmental lobbying by pressure groups, consumers and others

Table 1.1 Continued

Technological	• An ever-increasing use of social media, resulting in expectations of 'Instagrammable experiences' in aspects of the built environment (specifically retail and leisure) • The increase in online activism and the ability of a community to mobilise quickly over an issue of concern online • The increased use of proptech at all stages in development – from planning to sales and letting • The impact of technology on properties (Wi-Fi in the workplace, for example)

Notes

* The WELL Building Standard aims to revolutionise the way people think about buildings. It explores how design, operations and behaviours within buildings can be optimised to advance human health and wellbeing, centred around seven core concepts of health and hundreds of features: Air, Water, Nourishment, Light, Fitness, Comfort and Mind.

** www.wellcertified.com/certification/v2 [Accessed 15 September 2019].

male-dominated workforce that lacks diversity. Attempts to rectify the imbalance are showing signs of success,[14] but perceptions shift at a slower pace than statistics, and tackling sentiment (albeit with the help of statistics) is the PR challenge. Furthermore, the construction industry has relied heavily on skilled workers from the European Union, who, at the time of writing, are leaving the UK in greater numbers than they are arriving for the first time in a decade,[15] and more than a fifth of the workforce is over the age of 50.[16] Communications will be instrumental in recruiting and retaining talent into the industry.

Technological change

Proptech ('property technology') is having a substantial impact on the property industry. To name just a few examples, technology is today helping us to make better use of flexible space in buildings, to create a more service-based approach, to use sensors and 'big data' to monitor building use and performance, and to create digital twins and virtual replicas of buildings as tools to improve design, consultation and operation.

Technology in the sector will continue its rapid advancements. In 2019, PwC identified the 'Essential Eight'[17] core technologies likely to impact most on business across every industry over the ensuing three to five years: artificial intelligence, augmented reality, blockchain, drones, the internet of things, robotics, virtual reality and 3D printing.

In a global survey of CEO sentiment,[18] senior business leaders responded to this information: 76 per cent stated that they were worried about the speed of change and 64 per cent acknowledged that changes in the technology used to run their businesses would be disruptive. The survey concluded that emerging technology should be a key part of every company's corporate strategy. This advice applies equally to the property industry.

Appendix 1 (Technology across the property lifecycle) of the report *Lost in translation: How can real estate make the most of the proptech revolution?*[19] provides an excellent summary of the current opportunities for proptech:

- land identification;
- online land marketplace and site appraisal tools;
- consultant searching platforms;
- 3D visualisation and management tools;
- 3D model and geospatial databases;
- fintech solutions including crowdfunding/peer-to-peer lending platforms;
- market valuation apps and online mortgage brokers;
- technology to assist with sales and lettings, including remote viewing tools;
- location research platforms;
- tenancy management tools and rental passports;
- building monitoring/home management tools and technology to assist in access management.

Increased technology will invariably lead to greater collaboration throughout the property process. So interesting developments such as the emergency of 'cradle to grave' owner/manager/lettings companies common in Germany, where the rental experience is more closely aligned to lifestyle, is likely to become more commonplace in the UK via our burgeoning build to rent sector.

Additionally, the property industry is likely to be impacted by developments in autonomous vehicles and on-site power generation and storage. Surveyed by RICS in 2019,[20] 49 per cent of qualified surveyors stated that technology was having a 'moderate' impact on their role, while 34 per cent said that the impact was 'significant'. Looking to the future, 66 per cent felt that changing business models were likely to impact on their future role.

The benefits of proptech must be communicated both to the older demographic within the workforce – those who are accustomed to traditional ways of working and may, understandably, be unwilling to re-learn skills – and the emerging workforce, who will be entering a sector at odds with current perceptions.

The fact that PR is uniquely well placed to help the industry respond to change is evidenced in the British Property Federation's Technology and Innovation programme to support the sector in its digital transformation.[21] Each of its recommendations – from improving market information to championing innovation across the commercial property sector – is an initiative that requires communications support.

Changing communications

Technology is forcing change not only in property, but in PR too. As the following chapters will show, PR techniques have transformed over recent years.

Unquestionably the greatest change has been triggered by digital technology, which impacts on the way in which we communicate at every level. Many traditional PR tactics – such as press packs, media lunches, syndicated videos, wire services and direct mail – have now been replaced with online equivalents including webinars, augmented reality tours, blogs and vlogs, curation of information by algorithm, infographics, heat maps and search engine optimisation. Many offline tactics – from demonstrations to reviews and all other communications in between – now have a presence online too.

Online significantly increases the speed of communications and the amount of content available.[22] Some practitioners feel that that PR has become more reactive following the increase in requests for and expectations of information, while others claim that PR strategies must be increasingly proactive in order to get messages heard despite the distractions. Either way, the role of property PR increasingly involves curating, selecting and prioritising information to ensure that the target audience gains access to the most relevant content. The emphasis still needs to be on quality rather than quantity, and information such as research and crowdsourced comment requires greater interrogation than ever before.

Online, anyone can communicate with an infinite number of people, across the world, with zero budget, at the touch of a button. Inevitably, therefore, online communication has also given a greater voice to stakeholders, whose expectations of communication with companies have risen as a result.

The democratisation of communication has also resulted in a new emphasis on self-proclaimed experts, influencers and thought leaders and consequently, increased competition for a 'share of the voice'. According to research by Reputation.com,[23] the volume of online reviews has increased by a staggering 1,870 per cent since 2006, and consumers' trust in reviews has grown to the extent that in some areas of business online reviews are now on a par with word-of-mouth recommendations. Within the property industry, comment has increased in importance relative to statistics, partly due to the influencer effect, but also due to a rapidly changing political and economic scene, which renders five- or ten-year projections worthless.

The communication profession, rightly or wrongly, is responding to change by communicating more information, through more channels and more varied tactics, with greater speed. We are on constant standby to respond to immediate feedback and to be held to account by an increasingly broad and circumspect audience. There is no doubt that our role has increased.

Partly in response to the information overload, but also capitalising on new opportunities to do so, communication is increasingly targeted. While property companies now have the means to communicate directly with key audiences, they also have the opportunity to use influencers – whether internal or external – to target a message, with the benefit of an endorsement. We see this played out on social media, minute by minute. At the same time, property companies are increasingly using very specialist blogs to reach a clearly defined audience, and those that do this well ensure that their message is highly targeted to the interests of that niche group.

Ironically, perhaps, for an industry that values relationships highly, there is an increasing element of automation in targeting influencers, with online marketing agencies offering a service whereby content can be 'funnelled' to reach the ideal end user. As automation matures, the likely use of such methods to promote thought leadership will increase.

Structuring property communications

According to the Centre for Economics and Business Research,[24] 18 per cent of PR professionals work for consultancies, whereas 82 per cent are based in-house. Although a breakdown is not available for property PR specifically, it seems likely that the proportion of those working in consultancies will be greater, and this is reflected in the membership of the CIPR's Construction and Property Special Interest Group (CAPSIG).

There are many benefits to the property industry from opting to use specialist property PR consultancies. Morris and Goldsworthy (2016)[25] summarise the reasons for using a consultancy over an in-house team as follows:

- an understanding of regulation
- technical knowledge
- familiarity with the dedicated sector media
- size of business (small businesses tend to hire small PR companies).

With each of these features applying to the property industry, it is unsurprising that specialist consultancies are frequently used. And while some specialist consultancies service a broad range of property and construction clients, there are also niche consultancies focusing just on subsectors of the industry such as planning, retail or luxury residential.

The tendency towards specialisation is reflected in the CIPR's increasing number of special interest groups, including CAPSIG, which is one of the Institute's largest groups.

As the ways in which PR is structured are changing, so are the roles of the PR professional. In its PR and Communications Census 2019,[26] the Public Relations and Communications Association (PRCA) surveyed PR practitioners to identify the specific tasks that had recently increased in

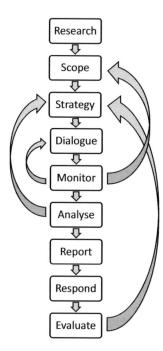

Figure 1.2 The strategic process.

importance. Unsurprisingly, 52 per cent believed that digital had increased most in importance, followed by online communication (28 per cent), search engine optimisation (SEO) (21 per cent), and reputation management (19 per cent). Those believed to have decreased in importance are sales promotion (29 per cent), writing articles/newsletters (19 per cent), general media relations (17 per cent) and event planning/organisation (16 per cent). While there are no figures available in relation to property PR specifically, it is likely that the sector reflects this change, and there is no doubt, as the following chapters show, that as we increasingly operate in an online sphere, the remit of the PR practitioner is expanding. The conclusion to this book considers whether in fact the term PR is too narrow for today's communications function.

The need for a strategic approach

While change impacts with seemingly increasing speed on every element of property communications, the importance of a strategic approach is a constant.

The property industry doesn't just require PR to promote its products and services, research, views and skills; respond to issues and crises; and deal with increasingly diverse audiences: it needs strategic PR.

At its most effective, PR is a strategic management discipline. Directors of PR and communications sit on the board of the top property companies, help to shape strategy to build brand value and enhance corporate reputation, and provide a bridge to the various other disciplines within those companies.

There are multiple models to show how PR strategies can be designed. But as Figure 1.2 shows, the most fundamental elements include using research to inform an approach; establishing underpinning aims for communication; clarifying messages; choosing the most appropriate specific, measurable, achievable, realistic and time-bound (SMART) objectives to achieve those aims; planning and budgeting; carrying out the approved programme; monitoring its success as it progresses; and using evaluation to determine its impact and to inform future improvements.

As this book will show, where a strategic approach is undertaken, messages are better formulated, audiences are better defined, work is better coordinated, and success is better evaluated. We hope that this collection of views about PR across a variety of property sectors will provide advice, guidance, inspiration and the opportunity for reflection.

Notes

1 www.savills.com/impacts/market-trends/8-things-you-need-to-know-about-the-value-of-global-real-estate.html [Accessed 15 September 2019].
2 www.ipf.org.uk/resourceLibrary/the-size-structure-of-the-uk-property-market-year-end-2017-update.html [Accessed 15 September 2019].
3 www.prca.org.uk/sites/default/files/PRCA_PR_Census_2019_v9-8-pdf%20%285%29.pdf [Accessed 15 September 2019].
4 The PRCA's PR and Communications Census 2019 (www.prca.org.uk/sites/default/files/PRCA_PR_Census_2019_v9-8-pdf%20%285%29.pdf) [Accessed 15 September 2019] classifies the market as follows:

- technology: 35 per cent
- consumer services: 22 per cent
- property/construction: 17 per cent
- health/pharmaceutical: 17 per cent
- business services: 16 per cent.

5 Norton, P. and Male, L. *Communicating Construction: Insight, Experience and Best Practice* (2020). Oxford: Routledge.
6 This chart originally appeared in BPF and Future Cities Catapult: www.bpf.org.uk/sites/default/files/resources/For%20web%20FINAL-%20Lost%20in%20Translation%20booklet_0.pdf [Accessed 8 September 2019].
7 www.theqca.com/article_assets/articledir_219/109701/QCABDOPULSEreport_Issue15_Autumn15.pdf [Accessed 15 September 2019].
8 A recent survey by the Home Builders Federation and the main warranty provider, NHBC (www.hbf.co.uk/policy/policy-and-wider-work-program/customer-satisfaction-survey-new), showed that 98 per cent of new home buyers reported snags or bigger defects to their housebuilder after moving in.
9 For example, the scandal over executive pay bonuses at homebuilder Persimmon: www.bbc.co.uk/news/business-46122407 [Accessed 17 October 2019].

10 www.parliament.uk/business/committees/committees-a-z/commons-select/housing-communities-and-local-government-committee/news/leasehold-reform-report-published-17-19/ [Accessed 17 October 2019].

11 MIPIM is the world's largest property conference but has come under fire following incidents of sexual harassment. Like many industry events, it is now taking action to tackle these problems and to promote greater equality, diversity and inclusion.

12 The criticism comes in the wake of the *Financial Times'* exposure (www.ft.com/content/075d679e-0033-11e8-9650-9c0ad2d7c5b5) of the sexual harassment and assault experienced by female hostesses at a men-only charity dinner hosted by the Presidents Club (https://thepresidentsclubcharity.co.uk).

13 Figures from the Intergovernmental Panel on Climate Change (2010), as quoted in www.propertyweek.com/legal-and-professional/is-the-property-industry-getting-serious-about-tackling-climate-change/5100244.article [Accessed 15 September 2019].

14 The RICS web page Diversity and Inclusion (www.rics.org/uk/about-rics/responsible-business/diversity-and-inclusion) provides updated information on the organisation's initiatives and success and increasing opportunities for LBGTQ+, women and other sections of society currently under-represented in the industry.

15 www.ons.gov.uk/peoplepopulationandcommunity/populationandmigration/internationalmigration/bulletins/migrationstatisticsquarterlyreport/august2019 [Accessed 6 September 2019].

16 www.prospects.ac.uk/jobs-and-work-experience/job-sectors/property-and-construction/overview-of-the-property-and-construction-industry [Accessed 6 September 2019].

17 www.pwc.com/essentialeight [Accessed 8 September 2019].

18 www.pwc.com/gx/en/ceo-agenda/ceosurvey/2019/gx [Accessed 8 September 2019].

19 BPF and Future Cities Catapult: www.bpf.org.uk/sites/default/files/resources/For%20web%20FINAL-%20Lost%20in%20Translation%20booklet_0.pdf [Accessed 8 September 2019].

20 www.rics.org/globalassets/rics-website/media/news/future-of-the-profession-post-consultation-report.pdf [Accessed 15 September 2019].

21 https://workplaceinsight.net/british-property-federation-announces-plans-to-modernise-commercial-property-sector [Accessed 15 September 2019].

22 It is estimated by Raconteur (www.raconteur.net/infographics/a-day-in-data) that by 2025, 463 exabytes of data will be sent each day – that's a staggering 43,000,000,000,000,000,000 bytes.

23 www.reputation.com/resources/case-study/property-management-industry-the-impact-of-online-reputation-management/ [Accessed 6 September 2019].

24 Research by the Centre for Economics and Business Research (CEBR) as quoted in Morris and Goldsworthy, *PR Today*.

25 Morris and Goldsworthy, *PR Today*.

26 www.prca.org.uk/sites/default/files/PRCA_PR_Census_2019_v9-8-pdf%20%285%29.pdf [Accessed 15 September 2019].

2 Promoting large-scale mixed-use schemes

Building a story for new communities

Emma Drake

Introduction

The planning, supply and ultimate quality of homes in the UK are probably among the biggest social issues of our time, and delivering the homes we need has been a political challenge for successive governments for decades. Not enough homes have been built, and in England we have demand for new homes far outweighing current supply. Part of the proposed solution to this challenge is to build large numbers of new homes in 'new communities', something that hasn't been adopted that often since the wave of New Towns in the 1950s–1970s. But this approach can bring large numbers of homes, jobs and facilities, and over the last decade there has been once again an increasing number of larger-scale new communities being both planned and built.

New communities are a unique type of development, quite unlike a regular housing development. They range from the smaller end, at about 1,000 homes, up to 5,000, 10,000 or even 15,000-plus homes. They can bring significant political, environmental and social challenges due to their large-scale and complex nature and the fact that they can take many, many years to complete.

The story or narrative of a new community can be caught up in the political cycle, meaning that they can suffer from policy and funding instability. These external disruptive factors make it tricky to tell a credible, consistent and sustainable story, and the inevitable change in project teams, project priorities and sometimes developers can also affect long-term relationships, reputation and trust with stakeholders.

However, the promise of new communities is also linked to the benefits they deliver – not just homes, but also new state-of-the-art schools, community facilities, new jobs, new transport links and sometimes major upgrades to railways, roads and public transport. The larger developments often include health provision, leisure facilities and new open spaces benefiting everyone in the surrounding area.

This chapter aims to cover key PR and communication approaches through the stages of building a new community, from supporting early-stage planning and consultation, site preparation and construction,[1] and 'place-making' through to selling a new destination.[2] It will focus on the purpose of PR and its role, the key areas of strategy, how this has changed over the years, and how it is perceived, as well as covering some of the challenges and opportunities for PR practitioners.

The context and purpose of PR

The resurgence of new communities provides an opportunity for a well-managed PR and communications programme: firstly, by building a powerful story of a new life somewhere fantastic with positive social out-comes, and secondly, by helping to differentiate competing places. The overarching ambition, however, is to manage the reputation of these devel-opments and to build and maintain trust with key stakeholders and the local community.

The challenge extends beyond the traditional remit of PR, as it aims to achieve a number of outcomes: balancing the interests of a variety of audi-ences, including potential residents and surrounding communities, political and other key stakeholders; delivering a compelling and believable story; building trust in long-term relationships; and ultimately supporting a suc-cessful commercial return for developers by supporting marketing, sales and future planning applications.

If we look back in history, we can see that story-telling has always played an important part. George Cadbury of the famous Cadbury's Chocolate had a vision for a new community for his workers, and in the late 1800s he relocated his factory and his workers from the smoke-filled slums of central Birmingham to the leafy new 'Bournville Village' to live in a brand new community.

George Cadbury was a philanthropist of his time, and his story is still being told today. He adopted a PR approach in promoting Bournville and the 'Garden Village' concept to the key influencers of the time, who saw it as a successful experiment in housing reform. It was at the Garden Cities Association Housing Conference of 1901,[3] which he hosted at Bournville, that the great and the good gathered and considered how housing was intrinsically linked to wellbeing, happiness and social mobility, and how industrialists and employers could do better by creating thriving, sustain-able new communities, like Bournville.

> 'No man ought to be condemned to live in a place where a rose cannot grow.'
>
> George Cadbury (1839–1922)[4]

Cadbury joined forces with the influential media of the day and used the *Municipal Journal* (ideally placed to reach his target audience) to launch the publicity around the ground-breaking conference. This campaign built the confidence and momentum needed, ultimately, to create the Garden City Movement and the design and construction of Garden Cities, initially Letchworth and then Welwyn Garden City. So, to a certain extent, PR played a key role in promoting property even in the 1800s and early 1900s.

From a theoretical perspective, this approach sits somewhere between the public information and two-way asymmetric models identified by Grunig and Hunt in their Four Models of Public Relations Communications.[5] The approach continued in popularity through the late 1940s following the New Towns Act (1946), which enabled a number of New Towns to be built up and down the country from the late 1940s through to the 1970s. PR teams 'sold the dream' of a new life in New Towns such as Stevenage, Milton Keynes, Telford, Harlow, Warrington and many others. Promotional videos, documentaries and extensive media coverage were used to promote a 'utopian dream' for families relocating for a better life outside London.[6]

Fast forward, and this approach was still being used effectively by housebuilders and developers through the 1980s, when marketing budgets were bigger, and the 1990s, when planning policy remained largely unchanged for some years.

But the most effective PR approach has shifted more towards the much-needed two-way symmetrical engagement model, partly as a result of the arrival of online communications and social media, and due to a lack of trust in institutions and politics and the economic downturn in the late 1990s.[7] The latter meant that, without the philanthropists, mutually beneficial partnerships and government intervention were needed to deliver new communities.

Today's consumers are vocal, well informed, and able to access information about corporate responsibility at their fingertips. They want more than

Table 2.1 Grunig and Hunt's four models of public relations communications

Characteristic	Press agentry/ publicity	Public information	Two-way asymmetric	Two-way symmetric
Purpose	Propaganda	Dissemination of information	Scientific persuasion	Mutual understanding
Nature of Communication	One Way: complete truth not essential	One-Way: truth important	Two-way: imbalanced effects	Two-way: balanced effects
Communication Model	Source-to-Resource	Source-to-Resource	Source-Resource-Feedback	Group-Group
Nature of Research	Little	Little	Formative: evaluative attitudes	Formative: evaluative of understanding

a glossy brochure with a front cover picture of a couple drinking cappuccinos. Health, wellbeing, sustainability and energy efficiency are becoming as important as location and functionality in a new community. This does not just apply to the new residents but to all stakeholders affected by its development. The PR and communications strategy has to navigate fierce competition and find the right story to tell; it must be unique, believable and authentic.

Some good examples of how two-way engagement is used when promoting a new destination can be taken from regeneration schemes, in particular when there is an existing community that needs to be brought along on the journey.

Box 2.1 Fish Island Hill Residential and Peabody: selling a new destination

Situated north of Bow in East London, Fish Island was formerly Tower Hamlets' largest industrial area.

It has undergone a transformation, earmarked for about 2,500 homes over the next ten years as part of a much wider gentrification of Hackney Wick, and attracts a thriving community of artists and designers.

Fish Island Village is at the heart, and includes a commercial space operated by social enterprise The Trampery, which hosts a strong artist community in low-cost workspaces.

The developers sought to work with the local community of artists to provide social harmony, to develop authenticity and to promote Fish Island as one of London's most exciting new places to live, with a unique history.

Development plans to gentrify the whole of Fish Island had been criticised in the past by the local populations, as it was assumed that regeneration would force out the local artists.

The team worked with local artists to create a destination brand, and developed key messages and content that underpinned both the brand narrative and the overarching proposition that summed up the local area. By taking a strategic approach, they were able to flesh out the real stories and passion, providing great campaign content. Local films used people's stories to add authenticity to the narrative and were used in digital and national press campaigns.

The team has fostered excellent working relationships with the local community in the process of developing the identity, and is confident that this will also be adopted by the rest of the regeneration area for that reason.

'We've had ongoing positive interactions with the local artistic community, including an artist taking residents and potential customers on a walking tour of the area, further local artistic commissions for marketing collateral, as well as print services derived locally.'

Matthew Clark, Key Property Marketing

Mind the gap: key challenges and how to overcome them

There are usually delays of some kind when building new communities. These can include campaigning by pressure groups and activists, delays to the construction programme, delays in planning approvals, economic downturns, uncovering of archaeological sites or environmental issues that need mitigation, changes in local and central government, building on or near Sites of Special Scientific Interest or historical or environmental significance, and changes in housing policy.

> 'There is no template. All sites are completely different, unique and distinct, with different issues, opportunities and audiences.'
> Rebecca Britton, Head of Communications, Urban and Civic

However, one challenge that is universally recognised is the need to have a strong enough story and proposition to keep the momentum going over a long period of time. The narrative has to grow, adapt and stand the test of time, as new communities can take years to come to fruition. Finding something interesting, engaging, authentic and credible to say on an ongoing basis is tricky, and the PR and communications programme cannot rely solely on news milestones, as these are few and far between. The story becomes easier to tell once the site is being built out and people are moving in, but it is much harder in the early stages of development.

The process of development lends itself to a variety of communication requirements. In the early stages, planning and site preparation, archaeological surveys, groundworks and earthworks provide opportunities and challenges, but consideration must also be given to potential gaps in communication. At this stage the main objective is forming relationships with key stakeholders and engagement with surrounding communities affected by the development: in other words, it's all about community relations.

Box 2.2 Eddington (North West Cambridge): creating a strong identity for a new district

The University of Cambridge has created a new district of about 3,500 homes to support the growth of the university and the city by providing affordable housing for university staff, student accommodation and housing for sale, alongside new local facilities and employment. The team wanted the district to have a unique identity relating to its history and heritage, so it engaged the local community to name the streets and the district. Wide-ranging engagement activity was undertaken through a Survey Monkey questionnaire and publicised in the local press and social media and through university channels.

Many people engaged with the campaign, and a number of names were chosen for the buildings and streets. One name stood out. Sir Arthur Eddington was a Cambridge graduate, an astronomer, mathematician and physicist.

He directed the Cambridge Observatory adjacent to the site and is buried nearby. 'Ton' is based on the Old English word *ton* (German root 'tun'), which means enclosure, estate or homestead, and is common in place names across the UK. The name Eddington was, therefore, chosen for the main spine road and, ultimately, the district name.

This successful campaign ultimately led to a genuinely authentic, unique and purposeful name, which became the 'destination on the front of the bus'. This process also provided lots to talk about and involved the community during a time when residents were moving in but the site was still under development.

The Eddington name also provided a strong basis for ongoing campaigns to reinforce the identity and embed the name both geographically and in people's minds:

What's your Eddington number?

This sustainable travel initiative noted the largest number of miles participants had cycled or walked for a consistent number of individual days. The campaign aimed to get more people cycling in the Cambridge area. Participants calculated their 'number' and tracked their progress using a campaign website and connected with other cyclists using Strava, a mobile app designed to connect runners and cyclists. This ultimately raised the profile of the name Eddington with a wider audience, helping to put it on the map.

Open Eddington

An annual event was developed to link into a wider city initiative (the Open Cambridge events) aimed at drawing the city out to Eddington, and also to embed the message that Eddington is 'open to all' as a new destination. Importantly, Open Eddington itself gave people a reason to visit and add Eddington to their mental map of Cambridge.

Both campaigns were successful in reaching and engaging with thousands of people over a two-year period.

> 'It's important in naming a new district that the "name on the front of the bus" resonates with people from the wider geographic area as well as with residents who live there. It has to have purpose and meaning for everyone to get behind it and get it on the map.'
> Heather Topel, Project Director North West Cambridge Development

Often sites have a 'ground-breaking ceremony', which involves a photocall with a shiny shovel and a local politician (even a minister in the case of a large-scale housing-led site). And then there can be a large gap, sometimes years, while the finer details of the planning and technical studies continue. Having the messaging and content to engage people through the next 18 months or more until buildings will be seen is crucial.

Box 2.3 Beaulieu Park by Countryside Zest (a partnership between Countryside Properties and L&Q) and the Land Trust: using strategic partnerships

Beaulieu is a vibrant new development of 3,600 new homes, schools, employment space and a neighbourhood centre; 176 of the 604 acres are estate parkland and open spaces.

The wide variety of green open spaces range from formal parks and pocket parks to community gardens and attractive street planting. There is also extensive tree planting – some 1,600 new standard trees and 30 acres (equivalent to 17 football pitches) of new woodland planting.

The unique green space and woodland proposition provided an opportunity to engage the community through a landscape-led approach and to promote a place that fosters health and wellbeing.

Countryside worked closely with the Land Trust as a strategic partner with Beaulieu Estate Management Ltd to take ownership of the green spaces. This was to safeguard and maintain the green spaces on Countryside's behalf, and to benefit the local community who would live there.

A programme of engagement activities to support and promote health and wellbeing was created around the open space and woodland planting, and a community engagement officer was appointed specifically to work with the early residents. This provided rich content for engaging with existing residents, and it also enabled a sense of community spirit and a value proposition that appealed to potential residents, who were looking to buy while large proportions of the site were still in planning or under construction.

As well as having practical benefits, this long-term approach to strategic engagement and maintaining a positive dialogue through each stage of development meant that Countryside maintained its good reputation through the various and ongoing phases of construction. It demonstrates how a genuinely unique proposition can positively affect engagement with early residents as well as potential homebuyers.

Building a core narrative that will communicate the reason why a new settlement is important and having a 'battle rhythm' that can appeal to a number of audiences over time are crucial. This requires data, insight, and creativity, understanding the unmet needs and utilising these to craft key messages that will resonate.

There is a huge amount of data freely available on housing numbers and buying habits through the Office of National Statistics, Census data, local housing strategy documents, government sources such as Gov.uk and charities such as Shelter. Carrying out primary research is costly but can deliver good, bespoke results.

By analysing different sources of data, cross referencing them and distilling this information, we can identify themes and unmet needs, which can be used to inform very specific and strong key messages.

Building the right messages to help reach the silent majority rather than just the sceptics can generate positive engagement with new audiences, specifically those interested in living in, or working in, the new development.

Box 2.4 Gilston Park by Places for People: using data to identify unmet needs and reach new audiences

Gilston Park Estate, based north of Harlow in East Hertfordshire, is an 8,500-home development site. It had been promoted by its former landowner as an urban extension to Harlow since 2004 but had attracted strong opposition, as a significant proportion of the site was in the green belt. Consequently, it was not allocated for housing in the East Hertfordshire District Plan. The backdrop was a lack of trust and engagement both within the local community and among local politicians. Places for People took over promoting the site in late 2008, initially through a joint venture with Land Securities and subsequently as the sole developer – acquiring the freehold of the 2,500 acres in 2011.

Consultation for a new 'estate' made up of villages began in 2009. Initially many local people and stakeholders refused to engage. To counter the charge that the development would be 'concreting over green belt' and that the existing villages would be subsumed within it, Places for People needed to demonstrate that new plans would be considerably different from the previous plans and developed a completely new masterplan concept for the site.

Re-engaging the existing, often negative, local audiences about this new plan was potentially challenging and required a new narrative and the benefits of the new masterplan to be communicated together with the benefit of large-scale development in this location and the urgent need for more homes in East Hertfordshire.

To commence the new approach, the name was changed to 'Gilston Park Estate', the historic name for the land.

Where will you live in 20 years' time? was launched in 2013. This opened up a conversation with local people about their experiences of the housing shortage in the area, which included a lack of affordable homes for young families, the inability to afford to rent (never mind to own), long commutes to work, and lack of high-quality choices for older people.

A core part of the strategic approach was a clearly identified set of messages to appeal to different audiences with varying degrees of awareness and knowledge. Data from community surveys, drop-in events and public consultations was used to understand local people's perceptions of the area and the improvements they would like to see.

The team set up meetings to discuss the plans in local pub gardens and on high streets in Harlow and Sawbridgeworth town centre, developed a 'prospectus' for key stakeholders, and took a 'consultation bus' around the local area – local media were urged to 'get on board with Gilston Park Estate Bus Tour'.

Work with local schools and colleges helped show a younger generation the opportunities that the development could bring to their future. Clever and engaging content utilising key messages was promoted via video and across social media, with the #YestoGPE targeted at potential first-time buyers/renters.

A new and refreshed identity breathed new life into the old project, a new website was launched, and an information pack was delivered to those within a 20 km radius. The target audience was directed to a website, which attracted over 275,000 unique users during the campaign period. Most importantly, face-to-face meetings were at the centre of the plans to enable two-way symmetrical engagement.

Following several years of intensive engagement, in 2018 the Gilston area was allocated within the East Hertfordshire District Plan for development of up to 10,000 homes. The outline planning application was submitted in May 2019.

Managing relationships: stakeholder engagement

Building a coherent vision for a new place, much like any new product or concept, takes time, effort and engagement from all parties. Stakeholders can include the county council (for schools and roads), the NHS and local primary care trust (for health provision), multiple housebuilders, Highways England, Homes England, Natural England, water and utilities companies, and local and special interest groups. It can be a cast of thousands.

Definition of a stakeholder

'Any group of individuals who can affect or is affected by the achievement of the [project/organisational] objectives.'
Edward Freeman, Author of *Management: A Stakeholder Approach*[8]

If we refer back to George Cadbury, it could be that he used the 1901 Housing Conference to influence his key stakeholders. He knew he had to rally support and created powerful key messages to tell his story and help others to follow in his footsteps to build better places for people to live in.

'There are very few quick-wins with communication approaches to large-scale development, you're in it for the long term.'
Ben Lee, PR manager, Bidwells

Things have moved on since the late 1800s, but by taking a strategic approach regarding stakeholder engagement, these partners can also become advocates and ambassadors for a new community, thus supporting the wider PR and communications programme, not only in a positive sense, but also tackling difficult issues. This approach involves identifying, prioritising and mapping out key stakeholders – including pressure groups and special interest groups, potential buyers and wider stakeholders. The Boston Matrix is a simple and effective tool to achieve this using the level of interest in the project and the level of power they have over its success or failure.

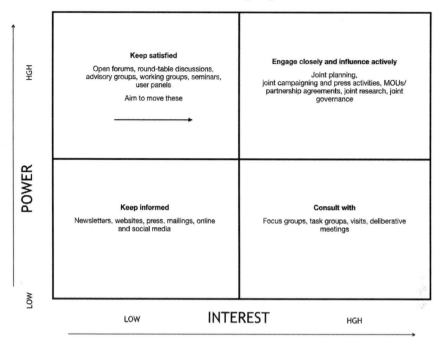

Figure 2.1 Boston Matrix for stakeholder engagement.

It was originally developed by the Boston Consulting Group in the 1970s to help identify business growth areas and has been adapted over the years to be used widely and effectively by practitioners to map and prioritise stakeholders.

Apathy is a common challenge when development sites have stalled and/ or communities have been over-consulted (sometimes referred to as 'consultation fatigue'). To help address this, stakeholders can also be prioritised in terms of their perceived 'closeness' to the project, and this can subsequently inform the sequencing of who to talk to first as well as the content and messages for communication and engagement activities.

Top Tip 1 Communications protocol

A communications protocol is a great tool for managing multiple stakeholders' needs and mitigating information leaks. It can foster a genuine two-way working model of communication between key stakeholders such as councils, developers, community organisations and funders. It helps set the parameters for joint working and establishes how communication should be carried out as well as agreeing sign-off procedures. Ideally, a cross-stakeholder communications group would steer the content, drive key messages forward and be collectively responsible for managing potential reputation issues.

In *Managing Public Relations* (1984),[9] Grunig and Hunt present the two-way symmetrical approach to communication as the one that best represents a truly professional approach to PR, stating that power 'is more evenly distributed between the organisation and its publics, with greater chances for mutual understanding'. While seen as idealistic by some practitioners, this definition epitomises good stakeholder engagement, and the approach can be very successful in building strong, meaningful long-term relationships in which power is balanced between a developer and the 'publics' or stakeholder groups.

Box 2.5 The Rise, Newcastle City Council: re-establishing a positive story post economic downturn

The Rise is a regeneration project in the West End of Newcastle being developed by partners Newcastle City Council, Keepmoat Homes and Barratt David Wilson Homes. The area had suffered due to the decline of heavy industries in the early 1970s–1990s.

This had caused worklessness and crime, social and economic decline, and empty homes – and had made the area unattractive, creating increasingly poor public perceptions of the wider area.

After several regeneration initiatives had stalled in the early 2000s, a joint public–private venture was established in March 2011 to deliver a £265 million housing-led regeneration initiative, designed to breathe new life into the area, kick start the local economy and revive the community.

This included creating about new 1,800 homes together with commercial and community facilities and investment in existing community assets such as parks and open spaces.

To tackle the embedded issues, to bring some momentum to the project and to create reassurance that the proposals would go ahead, Keepmoat Homes' Land and Partnerships Director worked with the local residents, through small working groups, on everything from the design of the buildings to the community facilities.

This gave the existing and surrounding local community a stake in its future success and contributed to a change in perceptions regarding the project, with people taking pride in the local area again, ultimately leading to the local community 'turning the earth' and significant amounts of positive local news coverage.

Reputation matters: stakeholders and issues management

The reputation of the developer and collective stakeholders involved in development is under constant threat from a number of environmental, societal, technological and economic forces. Latent publics can become active very quickly to assemble against a common cause, so building trusted groups of stakeholders and a common understanding is essential.

Unfortunately, PR and communications are not always seen as a strategic delivery function in this context, and this can be a challenge when issues and crises arise. Research supports this view: the *Handbook of Public Relations*[10] states that 'Public Relations is not commonly associated with owning stakeholder engagement' and that it [stakeholder engagement] is 'seen more as a management discipline'.

Using communication approaches to spot, monitor and mitigate issues before they become a crisis, however, is of significant benefit. Issues management is more than just a good media rebuttal. One leading crisis communications expert, Michael Bland, writing in *Communicating out of a Crisis*,[11] defines an issue simply as 'a crisis in a hurry', stating that it is the publicising of an issue or incident that makes it a crisis. He further describes issues as 'known' to an organisation, and crises as 'unknown'. Late identification of issues, therefore, invariably leads to a crisis and a firefighting task. Hainsworth and Meng's Issues Lifecycle explains how it is the publicising of the issue that intensifies the crisis, and that an issue can be mitigated up to that point.[12]

Top Tip 2 Issues logs

An issues log is a tool, sometimes in the form of a spreadsheet or table, that captures details such as the topic and nature of the issue; the stage it is at, and its perceived severity; and the mitigation measures in place should it unfold, which, depending on the stage the issue is at, can include actions that need taking, lines to take, internal briefing documents and key people to contact. This tool enables the PR team to capture, monitor and manage issues before they become a crisis and to share knowledge, analysis and insight of the issues with key stakeholders. Teams and organisations should develop a tool that works for them. However, there are some examples that can be used to inform such a tool – for example, Hainsworth and Meng's 'Issues Lifecycle',[13] which provides an example of how to rank issues according to the rising pressure and the four stages of development.

Often the PR and communications function will also be responsible for escalated customer complaints from residents, for example. Through an issues log, teams can keep one step ahead, monitoring potential and emerging complaints and feeding learnings into future phases.

> 'It's always better to build relationships in peacetime, as you'll need them when it all goes wrong.'
> Ben Lowndes, Director, Social Communications (Bristol) Limited

Of course, not all engagement is face to face. As with any PR or communications campaign, knowing how to communicate with audiences, including key stakeholders critical to the project, is paramount. Digital engagement

Box 2.6 Northstowe Town and Homes England: managing issues through stakeholder relationships

Northstowe, in Cambridgeshire, is a new town being built on a 240-hectare brownfield site and over 25 years. The site contained traces of chemicals and unexploded World War II bombs (UXOs), which can be commonplace on ex-MoD sites up and down the UK. Northstowe's PR team fielded an enquiry from a local journalist who had been made aware of this and wanted to publish a story about it. This was potentially very unsettling and worrying for those uninformed members of the public living in close proximity to the site.

The team had prepared an issues log and had worked closely with the UXO experts and a technical team to understand exactly how the work would be undertaken; the 'normal' or 'acceptable levels' of chemicals found on brownfield sites prior to remediation, and how the UXO disposal would take place safely. Expert responses were prepared, and the issues and mitigation measures were shared with key stakeholders closest to the project.

Because of the strong relationships in place, the team was able to work proactively with the local media and other stakeholders to manage the unfolding story and to provide significant technical detail and expert interviews, along with early stakeholder briefings. Community briefings were also arranged to coincide with the story's publication.

The journalist got their headline, but this approach to 'steal their thunder' took the wind out the sails of the story and provided the necessary factual content, thus preventing long-term reputational damage.

should be integral to the strategy, and the format works well in the long-term cycle of new communities, as it lends itself to growing an audience over time. As a project gains momentum, a small loyal audience can grow quite quickly, and interested parties' details can also be captured. People can follow events as they unfold, engage and ask questions, or clarify timings in between public consultations. Referring back to the Fish Island example (Box 2.1), audiences can share their own stories and ideas, and 'everyday influencers' can materialise, building trust and credibility in a project.

Top Tip 3 No 'no comment'

'Organisations are porous and misinformation can spread at a bewildering pace. Critics use this to their advantage, often posting inaccurate and unchecked content online. Those who issue a 'no comment' or instruct lawyers to shut websites down fail to address this problem. If you're faced with this issue, get your story straight and out into the open on your own terms. Signpost people to the right content and share it with stakeholders so they can do the same.'[14]

Ben Lowndes, Director, Social Communications (Bristol) Limited

Building a strong following online can also play a part in mitigating reputational issues as they arise by pointing an online audience to a web page for further information from a trusted source to counteract negative news. It can also, of course, be updated in real time.

What does success look like? Measurement and evaluation

'Awards are important for building credibility and reputation to deliver a site within the sector, but people don't move to a place because of how many awards it has won.'

Andrew Taylor, Director of Planning, Countryside

This section breaks down into two parts: first, best practice around measurement and evaluation, and second, defining what success looks like.

If we take the first, a common 'inputs, outputs and outcomes' framework can be used to measure the ongoing quantum and type of communications methods versus the effectiveness of activities in relation to the objectives set out in the strategy.

Key performance indicators (KPIs) for activities in this context can include:

- Increase/decrease sign-ups for newsletter by x per cent.
- Increase/decrease time spent on page x of website by x per cent.
- Increase/decrease shares by x per cent.
- Number of downloads/shares of x flyer for x event.
- Number of events tickets sold/attendance at events.
- Increase in 'opportunities to see' via footfall or reach data in specific geographies.

Effective evaluation frameworks include that of the Government Communications Service[15] – considered the 'gold standard' by many. Adaptations can easily be made to this to make it a bespoke tool. All activities, including community relations, marketing, sales and media relations, and stakeholder engagement, need to be monitored for outputs in order to inform an overall picture of outcomes. This requires working together with teams responsible for the various parts.

For digital channels, there are a number of free monitoring tools and resources available to measure reach and engagement, such as analytics from Twitter and LinkedIn as well as Google Analytics, Facebook, Tweetdeck and Mailchimp. RAJAR (Radio Joint Audience Research) can be used to assess radio reach and BARB (Broadcasting Audience Research Board) for broadcasting data.

Stakeholder engagement can be monitored through formal perception surveys or, more informally, simply by asking those who have responded to communications how it is going. There are also a number of automation tools available for online stakeholder management and engagement, including Trackability, Darzin and Kahootz.

But the true measure of a place is whether people like living there or not, and the quality of the residents' experiences. This brings us onto the second part. The level to which people feel happy, catered for and satisfied in a new community is hard to quantify. Established available 'measures' include the *Sunday Times* Best Places to Live rankings, which include quality of life, standard of accommodation, and access to jobs and schools. While from a PR perspective it is great to be on the list, this only considers established places. However, there is an opportunity for a communications programme to support measuring this outcome for new developments through setting benchmarks from research carried out at the start of activity or through surveys of ongoing perceptions.

In larger communities, there is a usually an element of long-term stewardship, a term given to the ongoing support a developer may provide as a 'custodian' or a 'master developer' overseeing the project, so there are elements, such as inclusion, happiness and satisfaction, that can be measured through residents' surveys over a period of time.

Other outcomes include something called 'social capital', which is tangible and can be measured by evaluating social connections and the benefits that they generate. There are four broad aspects of social capital: personal relationships, social network support, civic engagement, and trust and cooperative norms,[16] and some of these outcomes can be driven by community relations activities, such as the formation of local community groups, and via governance projects, for example.

Health, wellbeing and social purpose are also intrinsically linked to happiness[17] and are an important part of creating a sense of belonging to any place, especially a brand new one. These aspects are hugely relevant to new communities. Statistically, people with a good range and higher frequency of social contact report higher levels of life satisfaction and happiness and also better mental health. Interactions and relationships that support this fall largely into two categories: primary relationships, such as family and friends, neighbours and work colleagues; and 'strong ties' with family and friends. So, for new communities, there is an opportunity to measure social capital and focus the PR and engagement activity around health and happiness, such as communal and outdoor activities, including allotments and gardening or planting projects, or involving them in local governance. These, in turn, can feed into the evaluation of the project's success as well as supporting the sense of belonging to a new community.

Conclusion

New communities are a unique form of housing development. There is no one-size-fits-all PR strategy, and the core matters of reputation and trust remain as important as they do across many sectors. This type of development lends itself to long-term strategic planning and a communications strategy that is truly integrated across community engagement, stakeholder and relationship management, brand, marketing and sales.

Identifying objectives early, along with key audiences, and using data to inform choices on messaging and activities, utilising issues-management tactics and stopping to review progress over the course of the development, will ensure a robust plan and strong narrative that can stand the test of time. The PR team will need to be creative with content and work with wider teams, such as planning and community relations, marketing and sales, to develop interesting and believable stories and to ensure that the narrative remains strong and authentic. It takes a strategic approach and multiple skills to deliver this effectively.

The core skills required may change over time, but they are in essence people-led skills requiring human intervention, editing, sensitivity, emotional intelligence, applying good judgement and ethics, so automation of tasks through the introduction of AI (artificial intelligence) is unlikely to dominate. However, large parts of data capture and analysis are likely to have increasing automated elements to support this type of work, some of which, as previously mentioned, are already available. So, a PR professional's core skills will need regular updating to include a solid understanding of research, data analysis and data processing.

Finally, there is a huge responsibility in working with and promoting new communities. This can enable a real sense of social purpose as well as contributing to commercial returns, and is an opportunity to help shape something that many future generations will enjoy.

Notes

1 Planning and construction are covered in greater detail in the companion to this book, Norton, P. and Male, L. (2020) *Communicating Construction*, Oxford: Routledge.
2 Covered in greater detail in Chapter 4.
3 *Shaping Future Places, Bournville Garden Suburb: A Brief History*, Stride Treglown online, February 2018. https://media.stridetreglown.com/wp-content/uploads/2018/05/13181249/Shaping-Future-Places-Origins.pdf [Accessed 5 July 2019].
4 Ibid.
5 Grunig and Hunt in Windahl, S., Signitzers, B. and Olsen, T. (eds). *Using Communication Theory* (2000). London: Sage.
6 YouTube, 2019, New Towns: *Harlow, 1960's – Film 94050*. June 2019. www.youtube.com/watch?v=NNYvehXX8AM [Accessed June 2019].
7 Edelman, 2019. *Edelman Trust Barometer: Executive Summary*. www.edelman.com/sites/g/files/aatuss191/files/2019-01/2019_Edelman_Trust_Barometer_Executive_Summary.pdf [Accessed July 2019].
8 Stakeholder Theory online, unknown, *Management: A Stakeholder Approach*, unknown. http://stakeholdertheory.org [Accessed 5 July 2019].
9 Grunig and Hunt, *Managing Public Relations*.
10 De Bussy, N. 'Dialogue as a Basis for Stakeholder Engagement: Defining and Measuring the Core Competencies' (2010) in Heath, *The Handbook of Public Relations*.

11 Bland, *Communicating out of a Crisis*.
12 Hainsworth and Meng, 'Issues Lifecycle' in Regester and Larkin, *Risk Issues and Crisis Management in Public Relations*.
13 Ibid.
14 Lowndes, B. *Reputation Matters: How Developers Can Build Trust in Their Work* (2018). https://benlowndes.blog/2018/01/21/reputation-matters-how-developers-can-build-trust-in-their-work [Accessed 5 July 2019].
15 Government Communication Service online (2018), *Evaluation Framework 2.0*, civil service online, June 2018. https://gcs.civilservice.gov.uk/guidance/evaluation/ [Accessed 5 July 2019].
16 Scrivens, K. and Smith, C. 'Four Interpretations of Social Capital: An Agenda for Measurement' (2013) in *OECD Statistics Working Papers*, No. 2013/06, Paris: OECD Publishing.
17 *Measuring National Well-Being Quality of Life in the UK, 2018*, Office of National Statistics (2018). www.ons.gov.uk/releases/measuringnationalwellbeinglifeinthe ukapril2018 [Accessed 5 July 2019].

3 Promoting commercial real estate

Influence and innovation

Dan Innes

All companies working across the spectrum of commercial real estate – from developers, asset managers and investors through to advisors, proptech disruptors and placemaking specialists – have one thing in common right across the board – they truly value innovation and encourage creativity from their communications.

This investment in creativity and innovation in turn generates huge opportunities and revenue streams from a communications point of view. But increasingly we find that innovation doesn't stop at great schemes and powerful PR: it now permeates the way in which teams are structured, integrated and motivated.

As a result, we, as communications professionals, are increasingly able to have an input into business plans in the design and development process, influencing the product as never before. This lends itself directly to successful PR. It's a rewarding, virtuous cycle: the more novel the product, the greater the PR opportunities – and ultimately boosting the asset's value too.

It is often said that crisis inspires creativity, and this was undoubtedly the case during the 2008 financial crash. Those with experience of PR pre credit crunch will know how PR was once seen as a 'nice to have', often lacking a strategic focus and a clear means of evaluation.

Similarly, the property world was more complacent, less aware of the potential for values to drop without warning and unfamiliar with the rigorous performance metrics that exist today. The shock of the financial crash drove both PR and property to become more creative and more strategic, and to measure their results by exploring ideas such as evaluation using the Barcelona Principles 2.0 (described later).

Excitingly, this has triggered innovation, both in PR and property processes and in the resulting product, driving fast change in this retail property sector both in the UK and globally, and a huge surge in investment.

The years following the financial crash have also seen a much greater emphasis on the end user. Landlords need to engage closely and collaboratively with their tenants and work towards genuine partnership, just as those commercial and retail tenants need to engage with their own customers and provide an increasingly diverse range of services and experiences.

We know from many years' experience that good PR accelerates deal-flow and the leasing process, adding tangible value to leases to physical brands, portfolios and assets. When 50 per cent of Westfield Stratford City was sold to Dutch pension fund APG and the Canada Pension Plan Investment Board prior to its opening in 2010, the significant increase against original investment value took place 11 months before any shop had even opened and was largely attributable to market sentiment following a successful B2B communications campaign across the Westfield business. The timing, shortly after the 2008 financial crash and prior to any confirmation that the 2012 Olympics was to be held in East London, was counter-cyclical and not advantageous. But Innesco developed and drove key messages concerning the destination, positive learnings from the 2008 launch of Westfield London, and the new tenant mix offer at Stratford to reveal Westfield's resultant premier position in the London shopping centre hierarchy.

Just as greater integration between property and PR has triggered innovation, so has the broadening of the communications function. As Figure 3.1 shows, an integrated approach to strategic communications functions puts brand at the centre of the communications mix, and in doing so, both creates capacity for us to focus on the product itself and provides flexibility in communicating it.

Innovation in commercial real estate

The fact that the retail environment benefits from innovation is indisputable.

The 'perfect storm' facing our high streets and shopping places – brought about by a combination of rising costs, reduced customer spending, online shopping, out-of-town retail parks, and public transport and parking issues – has been the focus of many reviews, reports, conferences and committees.[1]

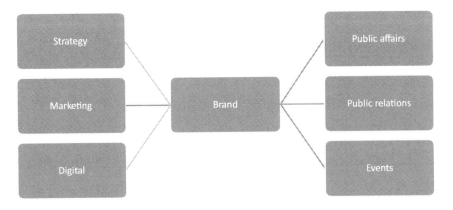

Figure 3.1 The Innesco approach to communications.

Without exception, the resulting recommendation is for innovation in branding, strategy, leasing and product mix, creating an offer bespoke to each situation, which can compete against the odds. Identikit malls are a 1980s anachronism. Psychology tells us that humans instinctively recognise similarities: it is change that captures their imagination and their appreciation.

'Shoppertainment' – a buzzword used by retailers to describe an overall in-store experience designed to bring customers back into brick-and-mortar stores – has been identified as part of the answer. Although retail overwhelmingly remains the primary driver behind shopping places, non-retail tenants – particularly those focused on culture, entertainment and leisure – are part of a fast-growing trend that is seeing a surge in spending power from millennials and families. This is exemplified in research that Innesco carried out recently (see Box 3.1), which shows that 70 per cent of those going to a cinema in a shopping centre said that they were much more likely to buy food before or after visiting the cinema, hence boosting dwell time.

In fact, leisure and entertainment occupiers are now increasingly anchor tenants of shopping places due to their impact on footfall, dwell time, sales and overall scheme attraction. Emerging leisure brands such as Flightclub and Puttshack have already taken huge spaces in centres such as Nova Victoria and Westfield London, respectively – to great success. This trend towards the 'wow' leisure and entertainment factor/spending power is epitomised by the forthcoming intu Costa del Sol project (slated to open in 2023), where, in addition to retail, the 175,000-square-metre site will include open-air clubs, a theme park, a wave pool, an aquarium, a year-round circus and a concert venue across eight differently themed and branded neighbourhoods.

Similarly, the number of health and wellbeing tenants has more than doubled recently, with fitness centres increasingly choosing to locate in shopping places, often linked to client demand, existing tenants, an established footfall and parking availability. One community-focused shopping centre, REIT – Capital & Regional – is actively supporting the collaborations between different tenants such as crèche operators and gym companies, helping customers to make the most of their precious time, now seen as life's most valuable commodity.

Food and beverage tenants – whether for a quick grocery shop, a snack from a pop-up restaurant, or a relaxing drink post work or workout – are natural stops for fitness health club users: so, as fitness centres increasingly position themselves in shopping places, the food and beverage offers are set to benefit. With the food and beverage sector also providing an important bridge between retail and other entertainment uses, the amount of space dedicated to food and beverage within existing European shopping places has grown to 20 per cent of all leases in the first half of 2018.[2]

Even the biggest chains are realising the central importance of becoming 'third spaces' (those dedicated to a use other than retail) to drive footfall and to capture the imagination and time of consumers. Complete sensory in-store experiences, such as in-house roasting and tasting sessions, basement cinemas, running clubs, membership vending machines and ice-cream parlours are now prolific, just to give an edge to a brand. Starbucks is a prime example of this, having unveiled its third Reserve Roastery in Manhattan: a sprawling outpost of sheer barista theatre housed within an incredible three-storey copper, concrete and wooden store design 'inspired by the history of manufacturing' in Manhattan's meatpacking district. The company has said it could open as many as 20 to 30 Roastery stores around the world.[3] The New York location is just one of six Roasteries expected to be open before 2020, following its first locations in Seattle, Shanghai and Milan – entirely a PR exercise.

Of course, digital also plays a huge role here, with 'Instagrammable experiences' becoming one of the biggest drivers of footfall in the modern age, and has comfortably taken its place in the pantheon of successful PR.

Entire cities can now be ranked by their popularity on Instagram. According to Big Seven Travel, the top five Instagrammed places in Europe are the Cotswolds, Budapest, the Highlands of Scotland, Iceland and in number one position – Dubrovnik. Quite possibly this could even be the reason why Soho House opened its Soho Farmhouse in the Cotswolds. With over one billion users worldwide, and as the second most engaged platform after Facebook, Instagram can make buildings famous – although the only modern building to make it into the top ten is the Burj Khalifa in Dubai (EMAAR), which enjoyed 1.7 million tags in 2017.

Increasingly, consumers' identities are formed by what they do, what they experience and what they share on social media, rather than what they own. The 2017 survey by Euromonitor reveals that, on average, over a third of European consumers prefer to spend on experiences rather than possessions.[4] These changing values are reflected by the leisure sector now attracting 150 per cent greater discretionary spend than retail and growing twice as fast: in the UK in 2017, 20 per cent of consumer expenditure was allocated to leisure.[5] The London and New York-based trend forecaster Stylus reports that excess digital consumption and acute stress are driving consumers towards real-life immersive experiences that stimulate multiple senses simultaneously and that trigger emotional responses.

We have seen this closely linked to the importance of sustainable consumption and the rise in importance of a circular economy when delivering major projects. The intu Costa del Sol project is to be built around a sustainable, circular economy and is intended to be one of the world's most sustainable and successful resorts, unlike anything seen elsewhere in the world.

These significant changes to the established, hackneyed tenant mix so criticised in the mid-Noughties as being 'homogeneous retail the same the world over' are now having a profound effect on how we communicate places and spaces today. Specifically, the changing attitudes of consumers, the experiences that inspire them and the more extensive use of shopping places that results from them present a multitude of opportunities to the PR and communications process.

Millennials, described as the 'experience generation', are now coming of age, and those in their mid/late twenties have high purchasing power. But increasingly, the 'silver generation' is also seen as having increased leisure spending power. A growing,[6] newly affluent, time-rich mature consumer is emerging, with relatively higher disposable income and a strong focus on health and wellbeing. Three-generational spending ('3-gen spend') is now a real phenomenon that landlords have tapped into, focused on a memorable family day out.

In response to the needs of all these consumer groups, and also in response to digital isolation, retail is going back to its roots as a true community gathering place – coming full circle to the birth of retail in the Agoras of Greece. Owners and managers such as Capital & Regional, Ellandi, New River Retail and Milligan are all embracing their own definitions of *community*, each with a higher percentage of non-discretionary retail tailored to local communities.

So, shopping places are continuously evolving, each creating a unique offering based on the priorities of the target market and creating a perfect combination of retail, leisure, food and beverage, community and other uses to combat competition.

Having once sounded the death knell for retail, online retail is now proven to depend upon bricks and mortar. CACI[7] estimates that 5 per cent of all online purchases in the UK are returned (compared with only 1 per cent of purchases made in-store) – and of these, 39 per cent of all returns are to a shop. Shops are showrooms, information providers, pick-up points, the physical representation of a brand and, of course, the only means a customer has of actually experiencing a product prior to purchase. Recent research by *Retail Week*[8] into the shopping habits of Generation Z, a generation that has never known a world without the internet, found that only 38 per cent of those surveyed preferred to shop solely online.

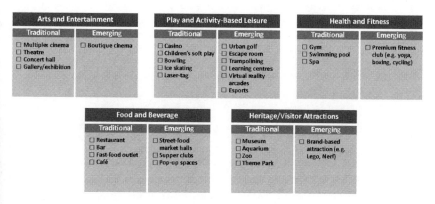

Figure 3.2 Examples of emerging and traditional retail occupiers (ICSC).[9]

So, while technology may pose a threat to retail property, it is also an opportunity. Technology has the capacity to bind the ever more disparate elements of the shopping centre offer and the potential to innovate further. And of course, technology is instrumental in communicating innovation.

With increasing links to retail, office development faces similar changes and challenges. In fact, the more innovative workspaces – co-working spaces such as WeWork, among others – are increasingly choosing to locate in shopping places, leasing ground-floor space to retailers and food and beverage tenants. WeWork is even extending into retail by setting up stores to sell its own branded merchandise and food. And, just as retail landlords have sought to make shopping places community gathering places, so too has community become a central pillar of the WeWork businesses.

Office developments are also increasingly partnering with providers of health and fitness in response to the wellbeing agenda.

Box 3.1 Chiswick Park: a case study in lifestyle-centric workplaces

Pioneering developments like Chiswick Park set the tone for the fusion of workspace and lifestyle space that today is commonplace but hasn't always been so. Chiswick Park was conceived as the antithesis of the traditional business park model, which originated in America but became the standard for (often, predominantly US) corporate occupiers in the 1980s and 1990s. These work environments sprang up during the early tech boom, centred on their proximity to airports, ample parking provision and modern, large floorplates. However, these somewhat practical attributes masked other failings such as a lack of public transportation and a lack of any leisure amenities, or even basic services such as ATMs or dry cleaning, all of which are essential to attracting today's workforce.

Chiswick Park was the first major development to tap into this changing trend. Despite having an out-of-town location and some of the hallmarks of the traditional office park, it differentiated itself by creating a brand – Enjoy-Work – that understood the needs of its end users: specifically, a changing workplace in which employers need to take into consideration the quality of environment that they provide.

The developers understood the need to provide not only modern, high spec office space for their occupiers, but also essential services (dry cleaning, banking, retail and leisure) and social interaction (events, sports clubs, social clubs and education/training) for the employees.

The end result has been both a communications success and a commercial success, with Chiswick Park having attracted occupiers including Paramount, Mitsubishi, Discovery, Sony and Pepsi and last trading at a value of some £780m. Operators like The Commercial Park Group are picking up this baton and looking at the next generation of office parks in newly identified hotspots for workforces, away from London's congested commuter belt and into fringe locations such as Bromley, Crawley and Haywards Heath.

Marcomms and real estate: an integrated approach

Communications can support innovation in real estate at every stage of the strategic process. The rationale is simple: positive sentiment grows because of innovation, and investment grows as a result of positive sentiment; and a project team imbued in the creative process is well placed to advise both on what forms of innovation can be communicated successfully and on how this can be promoted.

Research

Communications teams are ideally positioned to research market sentiment through both quantitative and qualitative methods, both formally and informally. By having a broad overview of a project and an understanding of a wide range of issues, from public affairs to social and business trends, communications professionals are well placed to carry out comprehensive and relevant research.

Box 3.2 Innesco/Maybe*: research to inform retail strategy

In 2018, Innesco collaborated with social media benchmarking service Maybe* on the largest social media listening project ever undertaken, to obtain and analyse consumer sentiment on the future for retail and entertainment. The findings, which were unveiled at MAPIC 2018, followed use of social listening, AI and chatbot technology and included the thoughts of more than 45,000 consumers.

The research contrasted shoppers' and retail professionals' opinions, highlighting key differences and similarities. Importantly, the research demonstrated that consumers' use of social media is at odds with that of most businesses.

Findings relevant to retail property and its promotion included:

- Consumers' requirements of physical shopping places included, in order of preference, outside seating 42 per cent; clean toilets 32 per cent; Wi-Fi 8 per cent; physical accessibility for those with additional physical needs 7 per cent and baby changing facilities 5 per cent.
- Eighty per cent said that a store would be more appealing if it offered food and drink stations throughout.
- Sixty-six per cent said that a special event would increase the likelihood of them going to a local shopping centre.
- Sixty-three per cent said that interactive art that brings people together would appeal.
- Fifty-five per cent felt that retail centres needed more Instagram-worthy places.
- Thirty-nine per cent of shoppers said that a cinema was the non-retail facility most likely to draw them to a shopping centre, and 21 per cent of potential cinema users were also likely to shop.

- Only 14 per cent of respondents claim to have had a store-based digital experience that has 'blown their socks off', which contrasts sharply with the number of retailers who claim to have provided such an experience.

Setting aims and objectives

The process of setting communications aims and objectives that respond to business objectives is something that all communications professionals are familiar with. Working with a real estate client on this, therefore, is a natural extension of what we already do well. Ultimately, asset managers' aims are growth in product value. Specific objectives may be to achieve this through leasing, brand equity, alternative means or a balance of each. Specific, measurable communications objectives are also best developed in parallel with this.

Messages

Communications teams, with expertise in writing and understanding communications theory and psychology, will shape the messages of an evolving business strategy. Knowledge of the industry and the specific project is necessary to gain a real understanding of the preconceptions, perceptions and limitations that will influence how messages are received. Both in the early stages and throughout the process, the communications team can liaise with all stakeholders – from B2B audiences, such as tenants, agents and the wider project team, through to the end user.

Audiences

Communications teams are experienced in carrying out stakeholder analysis to determine appropriate audiences and to gain an understanding of their relevance to the scheme, their priorities and their preferred means of communication. Effective communication results from awareness of the various influencers within the property sector and the relationships that are formed with those influencers. Compared with other sectors, this is still an emerging area in real estate. That said, editors and property correspondents of national newspapers are regarded as being of declining interest, whereas the respected analyst or commentator with an extensive LinkedIn network is in their ascendancy of relevance. Therefore, the ability of the communications consultant to identify the most appropriate influencers and analysts and ultimately to provide content that appeals to sentiment reports is of considerable value to a commercial real estate client.

Box 3.3 Cherrywood, Dublin: social media influencers

Innesco was appointed to oversee the B2B communications for Hines' Cherrywood Town Centre development, part of an entirely new neighbourhood in south-east Dublin.

For the purpose of the MAPIC Food conference in Milan, Innesco was commissioned to create a series of posts for the project's LinkedIn page. The posts were shared by the development director's personal account, whose extensive network resulted in the posts being seen by an average of 16,000 people: approximately twice the average net circulation of the dominant retail industry publication, *Retail Week*,[10] showing how the audience stage has tilted towards social over the last few years.

Strategy

Effectively setting or challenging strategy involves a combination of expert industry knowledge and familiarity with the creative process. Specifically, communications skills will be of benefit in shaping a unique selling point (USP).

Box 3.4 Creating a USP

Everyone at Innesco is inspired by the term 'outperformance', taken directly from the asset manager's lexicon. Not only does it imply excelling against the competition, but also 'getting out in order to perform' is what the team does well and is a practical tenet of good PR.

An in-depth knowledge of the professional sector can only be achieved though networking, much of it face to face, gaining first-hand feedback and sentiment. The communications team explores the nuances of a project to gain a full understanding of the market and comprehensively interrogate the leasing and asset management process to create USPs that are organically linked to a scheme's attributes. One such example would be the team's work in articulating Grosvenor Europe's unique 'Living Cities' approach, which combined a far-sighted, international perspective with an intimate local knowledge of markets and communities.

Box 3.5 Sale of EK East Kilbride Shopping Centre by Landsec: achieving value through innovation

Innesco was commissioned by Landsec (then Land Securities) to assist in raising the value of the recently extended shopping centre at East Kilbride. Our approach was to identify and publicise the innovative features within the shopping centre, to promote the leasing strategy and interest from retailers. As this was Scotland's largest undercover shopping destination, scale was critical to its messaging, and hence the team created an advertising campaign around the theme 'Big', which appeared consistently in key trade publications and formed the basis of direct marketing, media stories, site tours, a leasing lunch, briefings and interviews with analysts in both London and Glasgow. Once considered four separate assets that confused the market, our B2B campaign presented clear and concise messages of a single 1.3 million-square-foot scheme, boosted leasing, raised awareness, re-established its reputation, and drove a further 12 deals worth £1.5 million. The campaign won 100 per cent market recall. As a result, the scheme was then prepared for sale and sold to Propinvest in April 2007 for £380 million, representing a yield of 5.25 per cent, just months before the financial crisis.

Tactics and plan

As a developer rolls out an agreed strategy, the early involvement of the communications team enables the team to assist with the many communications aspects of the roll-out and to monitor its success. Just as retailers 'design for Instagram', the communications team encourages clients to 'design for PR'. This takes many forms, such as the name of the shopping centre, hashtagging its unique features, and the retailers and influencers that it seeks to attract. As part of this process, Innesco is frequently present at meetings between the client and the leasing agent, where we can advise on the catalyst deals that could substantially impact on the profile of the asset.

Box 3.6 Mood, Stockholm: a PR-led retail scheme

Innesco was appointed by the pension fund AMF Fastigheter, which had purchased the 1960s office block and home to H&M's global headquarters with a view to refurbishing it and creating a boutique shopping mall. We were instrumental from the early stages in advising on how to zone the space and naming and communicating the three very different spaces. We identified three 'character' zones – Everyday, Wishing and Dreaming. These characteristics informed the interior design, the tenant mix, the events and the promotion.

We also advised on the unusual strategy of signing restaurant tenants ahead of retailers. Due to the location of the new shopping centre (in the accounting/legal district of Stockholm and en route to the Metro), the new Michelin-star-equivalent restaurant provided the perfect catalyst deal to promote what was a unique and high-quality scheme.

As the development progressed, we worked with the client to introduce a variety of innovative features, from an extensive and opulent concierge to co-working chalets-on-stilts; from innovative fun features in the baby changing areas to public art designed to shock and be photographed for Instagram. The resulting tenant mix reflected the success of this bold approach.

Box 3.7 The Exchange, Ilford: adding community value

Innesco was appointed by Capital & Regional to oversee the promotion of strategic investment into its portfolio, which included The Exchange in Ilford. Capital & Regional was keen to develop closer links with the community, and Innesco promoted a series of initiatives centred around a new Family Zone, including a new toyshop, a parent-focused coffee shop and a new play area, which resulted in the shopping centre management forging stronger, long-term links with the local community and retailers.

Evaluation

When it comes to monitoring success, there are no better evaluation criteria for a communications consultancy than those set by the client, and therefore it is logical that the communications team and the client team jointly set evaluation criteria – which will be largely based on the objectives set out previously for the scheme's overall success.

Evaluation has become more difficult as communications consultancies and their clients utilise a wider variety of communications tactics. As asset managers/owners, Innesco's clients invariably evaluate success as return on investment: asset value, leasing enquiries or successful sale. But we also like to take a more scientific approach. We support the Barcelona Principles[11] and are reviewing Barcelona 2.0 with a view to adapting the principles for the property industry.

Until 2014, property companies had little hold on the success of their assets beyond parking, footfall and in some cases, revenue. At that time, most leases were 10–15 years and there was no obligation for retailers to release trading figures. As a result of several market forces, shopping centre operators now receive more frequent turnover figures, which assists in evaluating the success of the operation as a whole. Innesco's own proactive review of evaluation methodology has also proven links between PR, the rate of leases being signed and the value added to leases, alongside brand value and brand equity.

Integration and innovation

Increasing integration between developers and communications teams removes the need for a separate communications strategy: in a truly integrated client team, it is right that the communications strategy forms part and parcel of the overall strategy. There are many benefits in doing so. From the communications team's point of view, the crucial benefit is in having a formative role in the overall strategy or business plan. This provides unparalleled understanding of the strategic aims, as well as the opportunity to influence features that benefit communications. Furthermore, extensive knowledge of the project enables a communications team to convey genuine excitement in written materials and to write accurately about detailed aspects of the scheme.

A communications team can also assist a real estate client in overcoming restrictions on the creative process. While we have seen a considerable increase in innovation, there have also been additional restrictions placed on real estate clients. Changes in Financial Conduct Authority (FCA) legislation governing financial disclosures limit the ability to market a financial product including a property asset that is part or solely owned by a pension fund – as are most shopping places. Furthermore, the main legislation in the UK governing bribery and corruption is the Bribery Act 2010, which came into force on 1 July 2011 and can prohibit attendance at events unless the monetary value of the food and drink per head is supplied – which can pose difficulties, for example in the case of a buffet with a free bar, or tickets to a football match. But communications teams can find alternatives to standard corporate entertainment, simultaneously introducing more innovation and, therefore, interest in doing so. Similarly, recent General Data Protection Regulation (GDPR) legislation is a limitation, but one that can be compensated for with a strategic approach to social media.

Box 3.8 Westfield Stratford City: an integrated strategic approach

Innesco developed the key narrative for Westfield Stratford City as 'the Next Generation Retail' – emphasising key points of difference between it and the sister project in White City. Activities typically scratched deeper than ever before, liaising closely with brands' advertising and when they signed, as well as introducing several new platforms to the scheme – including the interview videos, site tours, helicopter photography, leasing newsletters and regular intensive media relations activity.

Creativity and communications are the most important attributes at each stage of the strategic process – whether in identifying new audiences and finding new ways of communicating with them or working with the client to introduce innovation to the product. It is this direct involvement in the design process that is undoubtedly the most valuable to the client, but it is also invaluable to the communications function, as such close links with the product provide the best means for us to promote it successfully.

The expansion of the communications role

PR is no longer separate from marketing, branding, digital communications, public affairs or event management, and as Figure 3.1 shows, an integrated service uses each of these functions as components of the brand.

Box 3.9 Malls by Tesco: creating a digital platform

Tesco's international property arm includes 400 assets in Eastern Europe, many of them shopping malls – where Tesco faced a communications challenge.

Evolving initially from an initial brief to create a hard copy display brochure for the CEO Dave Lewis, Innesco developed and programmed an app that enabled the Tesco team – from the chief executive to the asset and leasing managers – to present its 'Malls by Tesco' portfolio, allowing them to access the latest information about the extraordinarily large portfolio and to promote and present information with ease. The app enables relevant information to be selected and compiled, and a pdf to be collated and sent by email.

Innesco also ran a branding workshop and oversaw the promotion of the brand at MAPIC and other property industry events.

Such was the success of the digital platform for Eastern Europe that the same platform was extended to include the significant portfolio of centres in Thailand and Malaysia.

An example of the extension of communications is the renewed interest in study tours. The benefit of working internationally has enabled us to learn from some of the most innovative companies in commercial property and to share that knowledge. Frequently, we are requested to run study tours on behalf of clients who wish either to promote their work or to find out more from successful operators elsewhere.

Our most recent study tour was for the board of Swedish office owner-manager Humlegården, taking in tours with Google, British Land's new concept 'Storey', Broadgate Estates and WeWork.

Innovation in promotion

While innovation is indisputably the nectar of the promotional campaign, it is not always possible to work with a client to introduce innovation in the very early stages of a product's development. Good PR can be created by introducing innovation at a later stage, such as through marketing and events. Successful initiatives that we have observed in commercial real estate have included a variation of carpool karaoke, the popular segment from TV host James Corden, which was launched within a shopping centre as a means of recording a song on video that could then be shared on social media – both raising awareness of the centre and driving further traffic to the site.

In another example, a managing agent launched a competition for a new online-only brand to win a unit with a rent-free period and published the win by carrying their online customer base over the physical threshold. There are numerous examples of adding interest to stores by introducing lounge areas or pop-up collaborations to increase dwell time, or being innovative with visual merchandising to capture the customers' attention as they pass by.

Box 3.10 Slide into Summer: innovation

Innesco has managed numerous campaigns on behalf of Landsec at Trinity Leeds. Following its opening in 2013, Trinity Leeds has set a new benchmark for twenty-first-century shopping places, even winning the ICSC VIVA Best of the Best Award. It has a reputation as a launching pad for new retail innovations and leisure concepts, and a series of creative events are used to communicate this message.

'Slide into Summer' was an innovative campaign, which featured a helter-skelter running from the upper levels of the shopping centre into a changing room on the ground floor. The changing room, pulling in additional collaboration with local brands, provided a personal shopping service.

Conclusion

The interaction between PR and creativity in real estate from the Innesco point of view is described in Figure 3.3.

Put simply, the more creativity invested in the product, the greater its resultant value.

A successful product is the combination of the tenant mix, environment and services, technological advancements and branding. Increasingly, retail and office developers and managers can only remain relevant through an understanding of and regular communication with the consumer. Technology is integral to this, and although the pace of change is undoubtedly a challenge, it is also an opportunity.

Figure 3.3 The cycle of innovation and creativity.

Innovation is the currency and also the lifeblood of PR, and because we are fortunate enough to be able to influence the early stages of the development process – often three to five years before the launch or opening of a new project – we repeatedly see the benefit of this.

Not only does the opportunity to innovate drive good PR; it also inspires communications professionals and enables us to have a central role in the evolution of the real estate sector. As our social listening shows, marcomms and product development are increasingly linked, and this will continue, resulting in ever more innovative – and therefore valuable – future real estate, and consequently, greater respect for the communications profession.

Notes

1 The key texts being Swinney, P. and Sivaev, D. *Beyond the High Street: Why Our City Centres Really Matter* (2013), London: Centre for Cities; Breach, A. and McDonald, R. *Building Blocks: the Role of Commercial Space in Local Industrial Strategies* (2018), London: Centre for Cities; *Future High Street Fund Policy Paper* (2018), MHCLG; *High Streets and Town Centres in 2030* (2019), London: Housing, Communities and Local Government Committee; *Revitalising Town Centres* (2018), London: Local Government Association; Grimsey, B. *The Grimsey Review 2* (2018); Grimsey, B. *The Grimsey Review: An Alternative Future for the High Street* (2013), http://peyeprod.wpengine.com/paymenteye/wp-content/uploads/sites/19/2014/01/GrimseyReview04.092.pdf; Timpson, J. *The High Street Report* (2019), MHCLG; Portas, M. *The Portas Review: An Independent Review into the Future of Our High Streets* (2011), https://assets.publishing.service.gov.uk/government/uploads/system/uploads/attachment_data/file/6292/2081646.pdf.
2 CBRE, 'EMEA Real Estate Market Outlook 2019' November 2018. As quoted in www.icsc.com/uploads/t07-subpage/European_Trends_Industry_Sector_Series.pdf [Accessed 15 September 2019].
3 As quoted in www.cnbc.com/2018/12/12/heres-what-starbucks-new-roastery-in-new-york-city-looks-like.html [Accessed 14 October 2019].
4 Euromonitor International Lifestyles Survey 2017. As quoted in www.icsc.com/uploads/t07-subpage/European_Trends_Industry_Sector_Series.pdf [Accessed 15 September 2019].
5 Ibid.
6 According to Coresight Research, more than one-quarter of the total population in western and southern Europe will be aged 65 or older by 2030. Coresight Research (formerly known as Fung Global Retail & Technology), 'Mining Silver: Identifying Opportunities in the Senior Boom', 2017. As quoted in www.icsc.com/uploads/t07-subpage/European_Trends_Industry_Sector_Series.pdf [Accessed 15 September 2019].
7 As quoted in www.icsc.com/uploads/t07-subpage/European_Trends_Industry_Sector_Series.pdf [Accessed 15 September 2019].
8 www.revocommunity.org/blog/appeal_real [Accessed 15 September 2019].
9 www.icsc.com/uploads/t07-subpage/Leisure_Location_BasedEntertainment_Industry_Sector_Series.pdf [Accessed 15 September 2019].
10 https://en.wikipedia.org/wiki/Retail_Week [Accessed 15 September 2019].
11 The Barcelona Principles are a set of seven principles that provide an overarching framework for communications measurement. The principles were originally adopted by about 200 delegates from over 30 countries at the Second Annual European Summit on Measurement convened by the International Association

for Measurement and Evaluation of Communication (AMEC) in Barcelona in 2010. The principles were developed with, and supported by, AMEC, the Global Alliance, the Institute for Public Relations, the International Communications Consultancy Organization, the Public Relations Consultants Association and the Public Relations Society of America.

- Principle 1: Goal setting and measurement are fundamental to communication and public relations.
- Principle 2: Measuring communication outcomes is recommended versus only measuring outputs.
- Principle 3: The effect on organisational performance can and should be measured where possible.
- Principle 4: Measurement and evaluation require both qualitative and quantitative methods.
- Principle 5: AVEs are not the value of communication.
- Principle 6: Social media can and should be measured consistently with other media channels.

The Barcelona Principles outline the basic principles of PR and communication measurement and represent an industry-wide consensus on this topic. They are intended not only to demonstrate proof of performance but to show how to foster continuous improvement. The Principles serve as a guide for practitioners to incorporate the ever-expanding media landscape into a transparent, reliable and consistent framework. They are considered foundational in that specific measurement programmes with clearly stated goals can be developed from them.

The Barcelona Principles identify the importance of goal setting, the need for outcomes, instead of outputs-based measurement of PR campaigns, the exclusion of ad value equivalency metrics, the validity of quantitative and qualitative measurements, the value of social media, and a holistic approach to measurement and evaluation. Each principle highlights many of the quantitative and/or qualitative approaches practitioners can follow and also accepted methodologies to put these principles into practice.

The Principles are relevant to organisations, governments, companies and brands globally. While initially adopted in June 2010, they have now been updated to reflect changes in the communication field with input from a wide array of organisations and individuals. As next steps, the industry needs to support the continued adoption of the principles and help professionals understand how to apply them. Also, regular reviews to reflect changes in communication should be done every three to five years as relevant. Adapted from https://amecorg.com/barcelona-principles-2-0/ [Accessed 15 September 2019].

4 Promoting homebuilding

Positive communication in delivering new homes

Clare Jones

Introduction

Managing communications around homebuilding is one of the most fulfilling yet challenging areas of PR and communications within the built environment. This is not just about bricks and mortar. We are talking about people's homes, promoting a vision and creating communities in which people want to live.

Buying a new home is a major investment, and people want a property that will meet their lifestyle aspirations, now and in the future. Simply developing well-designed properties is not enough. People's priorities are changing, and many want to be part of sustainable communities with facilities and amenities on the doorstep as well as good links to the wider area.

On the surface, ensuring positive PR around new homes should be straightforward. We are in the midst of an ongoing housing crisis, with the demand for new homes far outstripping the rate of delivery; resulting in inflated housing prices and high levels of homelessness. Surely ensuring local support, interest and buy-in for a new housing development should be a given?

But as anyone involved in the sector knows, this discussion is framed by a range of issues and emotions that create what can often be a problematic tension when communicating around the development of new homes.

There are a number of contributing factors, including regular strong opposition to the principle of new housing development (in particular in relation to greenfield sites), a general misconception from local communities that housing is not needed in their area and should be provided elsewhere, and concerns about pressure on existing services and infrastructure. In addition, the backdrop of negative press coverage around some housebuilders has brought into question the quality of new housing schemes and the ethics of operators in this sector.

When exploring PR for homebuilding, we need to consider the management of communications at a local level on a project basis to ensure that schemes gain planning permission and can be successfully marketed to potential buyers, while also bringing positive benefits to the local community. It is also important to look at the wider national reputation management of the homebuilding sector and how the impacts of this need to be considered and managed at a local level.

There is an unavoidable interplay between project communications locally and the impact this can have on the wider reputation of the home-builder (both positive and negative). Corporate reputation can also have a huge impact on perceptions of a specific project. For example, the Home-Owners Alliance 'Building Better Homes' campaign is lobbying for better-quality homes.[1] It is working with unhappy new home owners who have experienced difficulties resolving problems with their new builds, and 'Regularly highlighted members stories in the media (with member's consent), often encouraging developers to resolve the problems.'[2]

This shows how quickly a local issue can hit the national news, creating headlines such as:

Is there a crisis of quality in new-build homes?

Guardian, 17 November 2018[3]

Homeowners £229,000 new-build dream house turns into nightmare with long list of snags.

Mirror, 22 April 2018[4]

This chapter will explore the wider context and issues that frame communications around homebuilding, look at PR requirements and objectives throughout the cycle of a typical homebuilding project (from pre-planning engagement through to successful determination, construction and selling new homes), and explore in more detail some of the tools and tactics that can be used to help meet these communications objectives.

National context framing local strategies

Industry reputation

There are hundreds of developments coming forward every year across the UK – in a wide range of sizes, designs and prices – with varying degrees of community and stakeholder engagement and proactive PR surrounding them.

The 2018 annual National New Homes Customer Satisfaction Survey (CSS) carried out by the Home Builders' Federation (HBF) and the NHBC, which was issued to almost 100,000 buyers of new build homes, paints a highly positive picture about the quality of new homes being built:

- Eighty-seven per cent of new buyers would 'recommend their builder to a friend', the key question on which the HBF star ratings are based. This is up from 86 per cent in 2017 and 84 per cent in 2016.
- Over 90 per cent of those who bought a new build home would buy new again.

- Eighty-six per cent were satisfied with the quality of their new home, up 1 per cent on the previous year (5 per cent neither satisfied nor unsatisfied).[5]

However, the CSS is itself coming under some scrutiny – it only applies to those HBF member builders who wish to take part, it obviously only reflects the views of those home owners who choose to respond (about 60 per cent of the total sample), and it only gets sent to purchasers approximately eight weeks after legal completion, so they are still very new in their homes. Changes to the CSS are being considered for the next survey.

It also only takes a couple of negative local or national headlines about the homebuilding industry, or about a specific developer or scheme, to disproportionately influence public opinion and to create negative perceptions that can then be difficult to change. This can also make future positive engagement and communications a real challenge.

Senior influencers can have a huge impact on perception and reputation. The quality of the homebuilding industry has recently been put in the spotlight as a result of direct criticism by senior government and political figures.

A particularly stark example of this is Welsh Government Minister for Housing and Local Government, Julie James AM, who featured prominently in the Welsh headlines after accusing the housebuilding industry of 'building the slums of the future'.[6] This is an example of how damaging senior stakeholder and political comments can affect an overall industry – in this case, by promoting a public perception that all or most housebuilders are building poor-quality homes that fail to establish high-quality places and communities. This makes it more challenging for homebuilders to establish relationships at a local level when developing new schemes, as well as potentially when marketing these new homes after they have been built.

Use of language here is key. This is demonstrated by comparing the negative comments of Julie James AM with those made at a UK Westminster level by the then Minister for Housing, Kit Malthouse MP, who made a statement requesting that housebuilders strive to improve the design of their schemes: 'I tell developers to bear two things in mind: they're not just building homes but building neighbourhoods; and they need to ask themselves whether they're building the conservation areas of the future.'[7]

It is clear that when it comes to reputation and managing PR for homebuilding, a wide-ranging strategy needs to be considered that encompasses media relations and proactive communications and messaging through online content and social media channels, but that also considers the best way to communicate with key influencers and stakeholders, be it at a local, regional or national level. Garnering support from politicians, especially at a local level, can often be very challenging when it puts elected members in direct opposition to the very constituents on whom they rely for re-election.

Housebuilder headlines: PR and policy issues

Aside from comments from senior influencers calling into question the quality and ambition of schemes being built by major housebuilders, there have also been a range of other policy changes and news stories that continue to frame messaging and dialogue around homebuilding in the UK. These need to be considered when planning a communications strategy to help identify risks and address potential issues.

Examples of these include:

- **The housing shortage:** research carried out in 2018 indicates that England has a backlog of 3.91 million homes, meaning that 340,000 new homes must be built each year until 2031. This figure is significantly higher than the government's current target of 300,000 homes annually.[8]
- **Help to Buy:** controversy around the efficacy and profit-inflating impact of the UK government's Help to Buy scheme.[9]
- **Executive bonuses:** there has been extensive negative press around senior executive bonuses, resulting in extensive media coverage, criticism and public anger.[10]
- **Leasehold row:** stories around new home owners feeling they have been 'mis-sold' unfair leasehold contracts – resulting in headlines such as 'We've been caught in a leasehold trap' (BBC News, 12 June 2019).[11]
- **Social and affordable housing:** the interplay between the delivery of private sector housing and social or affordable housing.

These headline examples all serve to create background noise that impacts on perceptions of homebuilding and the construction industry in the UK. It is important to understand these issues before seeking to engage with local communities on homebuilding.

Strategic approach: setting the scene, localising the message

Reputation management

It is easy to become focused on the PR surrounding the major players in the homebuilding market, partly because of their increased dominance in recent years and the decline of smaller companies. According to a 2017 HBF report, only around 12 per cent of new homes are delivered by small and medium-sized enterprises (SMEs).[12]

However, there are plenty of opportunities to tell good news stories within the industry, including about the quality of new homes being built, customer satisfaction of new home owners, and the innovation and ideas coming forward in design and construction. This requires strategic communications campaigns by both individual homebuilders and, over the longer term, the industry as a whole.

Arguably, more homebuilders need to be bolder about the positive work they are doing, not only in terms of their product (such as build quality, design and landscape setting) but also around the way they operate (such as jobs and skills development, CSR activities and sustainability credentials) and socio-economic benefits that homebuilders provide (such as enabling the delivery of social housing and training and employment opportunities for local people). The message for each company needs to be clear and distinctive so that not all developers are tarred with the same brush.

Therefore, when planning a PR and communications campaign, as well as considering the wider reputational context, it is important to understand the local context of a project and to identify the key unique and positive features of the project. Undertaking a SWOT analysis (strengths, weaknesses, opportunities and threats) and a PESTLE analysis (political, economic, social, technological, legal and environmental) on the project and on relevant local, regional and national issues will help inform the communications strategy and campaigns, shaping the messaging used and helping the team prepare to address potential issues. In addition, undertaking an audience and stakeholder analysis will further help shape the strategy and key messaging.

There are a variety of developers building a wide range of different types and styles of homes, targeted at different audiences, in different parts of the country. Communicating around homebuilding is about much more than just individual homes; it is about creating new neighbourhoods or contributing to the expansion of existing communities and ensuring access to the necessary social and physical infrastructure.

Clearly, the approach for large housebuilders will be different from that for smaller developers due to a wider corporate and national PR context. This provides both opportunities (such as track record, efficient delivery and CSR approaches) as well as risks (in relation to negative PR or customer feedback).

However, for all projects, regardless of scale or specification, identifying the key opportunities and likely issues early on is an important place to start.

Capturing the brand and vision, creating positive profile

While working with a design team to understand the plans, and getting under the skin of the brand and product to understand what makes it unique, understanding the detail is key to creating a compelling campaign.

This ultimately is the key to delivering great PR, by tailoring key messaging to identify and articulate what is unique, exciting and positive about a new scheme, to help capture media attention and to generate positive messaging using a range of channels.

Developers are increasingly offering homes with unique and dynamic specifications, such as low carbon homes or luxury living, which clearly offer wide scope for creating interesting content and media coverage. There are also plenty of examples of innovative features and CSR approaches being incorporated into more conventional schemes, which equally need to be shouted about to help raise the overall profile of the industry.

Box 4.1 Sero Homes: capturing brand through community engagement

Industry newcomer Sero Homes is committed to reflecting its sustainable, high-quality product throughout its brand, communications and community engagement approach. When promoting the public consultation on proposals for 35 zero-carbon homes in Pontardawe, South Wales, Sero Homes sent invitations to local residents on seed paper. As well as information about the project and engagement approach, the invitations included instructions on how to plant the paper to grow wild flowers. This reflects the integrated approach to landscape and ecology that is included in the plans for the new homes, which are designed around a central shared green communal space with extensive landscaping and planting throughout the scheme.

The public exhibition materials were also printed on recycled coffee cup paper. With future generations in mind, specific engagement materials were also produced for children, including a poem about the project and a fold-out colouring booklet.

Delivering new neighbourhoods

Compared with smaller developments, it is often easier to package the benefits of larger residential-led developments into positive messaging about the scheme, due to the range of opportunities (such as new schools, community centres, managed green open spaces and play areas, etc.), and to demonstrate that a masterplan provides the blueprint for the delivery of a new cohesive neighbourhood. It is also easier to create a distinct identity and to articulate an end vision for this type of scheme, providing strong hooks for delivering media coverage and compelling content for online campaigns.

However, this can also present challenges, as early on in the project it is necessary to sell a vision rather than a reality. People will be buying a home that is part of a building site, with actual facilities and amenities not yet in place. It is therefore important to manage expectations and to make sure future residents understand the phasing and timing of the delivery of the overall project. For further information and for case study examples in relation to major urban extensions and masterplanning, please refer to Chapter 2.

It is perhaps more challenging when planning strategic messaging for smaller projects that will not deliver the same sort of community amenities or bring enough revenue to deliver the types of highways improvements or new facilities that existing residents believe are needed to support the additional homes proposed. In these cases, the role of professional PR can be to help identify and shape a small scheme to become more responsive to local need, or at the very least to be seen to be listening to, and mitigating, local concerns.

Objectives, evaluation and re-aligning your strategy

Setting out measurable communication aims and objectives at the start of a campaign is key, as this will inform the overall strategy and help prioritise the tools and techniques to be used.

It is also important to establish a system of ongoing monitoring, with relevant metrics in place, to ensure that PR and communications activity is meeting the identified objectives and that opportunities are being appropriately maximised. All strategies and campaigns require a degree of flexibility. So, ongoing evaluation allows the PR team to understand when to drop less effective activities, or to build up successful ones, in order to maximise results.

Communicating the vision: planning communications

Proposals for new homes are often (although not always) controversial locally. Proactive engagement and communications are essential in ensuring that a project is launched on the best possible footing.

In England, there has been significant change in the planning system since the Regional Spatial Strategies were abolished, which has, in general, made local communities feel more empowered, active and aware about planning and development. In addition, the Localism Act placed an increased requirement on the need for engagement with communities throughout the planning processes, neighbourhood plans have provided an opportunity for local parish councils to feel empowered to effect change, and Local Plans have required cooperation between local authorities in a whole new way.

Engagement at the earliest point is important to demonstrate openness and transparency, but if this early communications programme is not carefully managed, it can result in inflated levels of local concern and mistrust, fuelling potential opposition to the proposals, which can contribute to delays in planning and delivery. The local rumour mill can build momentum very quickly, and it is important to ensure that accurate information is circulated and that factual inaccuracies are addressed.

At this early stage of the project, a comprehensive strategy needs to be put in place that includes:

- Stakeholder mapping to identify the target audience – including influencers, neighbouring communities and potential buyers – and their likely interest and influence in the project.

- A SWOT analysis to identify likely local concerns and aspirations and project opportunities for the communications campaign, which can be fed into key messaging and FAQs.
- A programme of communications activity based around the project milestones.
- Establishing objectives for the communication strategy. These can include: delivery of a compliant consultation programme in line with relevant local/national policy and guidance; engaging with the silent majority (many of whom may be project supporters); ensuring a proactive response to issues raised on public forums; and identifying and engaging potential future buyers for the homes.
- Agreeing communication protocols with the client and project team to help save time and to ensure that issues and queries are responded to efficiently as the project progresses.

Responding to queries in a timely manner and addressing misconceptions are important when managing reputation and building trust at a local level. Even if a resident or local member doesn't like the answer, it is still important that they can get an answer. While inevitably there will be some criticism of the proposal, the consultation and engagement should be of such a standard that it can be robustly defended if it becomes the subject of the criticism. This can be made much easier by having agreed processes and sign-off procedures in places, as well as holding statements for the media. Also, ensuring that a meaningful frequently asked questions (FAQ) document is put in place will help ensure that the project communications can be managed from the initial announcement of the plans, through construction, to the first residents moving in.

There are a number of key issues that are invariably raised during pre-application consultations for new housing developments that can be anticipated and prepared for in advance. Examples include:

- traffic, increased local congestion at peak times and capacity on the road network;
- access (or lack of access) to public transport;
- amount of off-road parking and impact on existing residents;
- levels and impact of social housing to be included within the development;
- need for open green spaces and play areas;
- impact on existing ecology and opportunities to increase biodiversity;
- impact on existing heritage or archaeological assets;
- construction impacts and management;
- impact on existing facilities and amenities – including schools, doctors and dentists;
- potential flooding – including surface water and fluvial;
- foul drainage capacity;

- potential benefits – including supply chain opportunities during construction.

The initial research previously referred to can also help identify site- or community-specific issues, which can then be prepared for ahead of public engagement.

Following the consultation process, it is important to ensure that a response from the developer is provided to complete the engagement circle. If people have taken the time to review and comment on proposals, it is important to acknowledge their input and to address key issues raised; explaining how they have been responded to – where changes have been incorporated or, if they haven't, why not. Ongoing relationship building with stakeholders and communities is an important part of any homebuilding PR strategy.

Reframing the debate

One of the keys to successful local engagement around developing new homes is how the dialogue is framed. It can be difficult to move away from a debate over the principle of development to a wider discussion about the design and opportunities of the scheme. In many cases, developers are taking forward plans on sites that have been allocated for housing within a Local Plan, meaning that the debate over the principle of development has already taken place and been resolved. However, local residents are not always familiar with the Local Plan policy process and may not have engaged in a Plan process, which may have happened some time ago, and so end up feeling frustrated from the outset. This issue around principle of development is often one of the first communication hurdles that need to be overcome.

Large residential schemes will have scope and viability to incorporate elements that benefit existing and future communities, but a key challenge can be to establish the scope of consultation so that communities understand what can really be influenced. However, the use of early research and engagement with local representatives can provide an early steer on the issues that will capture the interest of local residents and stakeholders. Financial constraints will always be a consideration, but close working with the project team and creative thinking at early stages in the design of the scheme can result in real changes that genuinely bring about community benefit. For example, there is often an interest in the inclusion of space for a community food growing project, or requests for a particular size or type of housing that is scarce within the local area (such as homes suitable for older or disabled residents).

Other community benefits that local residents are likely to look for through a new housing scheme (depending on the scale) that range in cost and deliverability include:

- family homes with gardens
- homes for first-time buyers
- accessible homes for disabled or older residents
- improved footpaths and cycle links within a locality and connecting to the wider network
- improved public transport services
- improvements to the local road network
- new community facilities, such as community spaces, shops, health services, nurseries and schools
- open green space or play areas
- allotments or community garden spaces
- improved broadband
- sustainable drainage systems that help address existing issues
- appropriate redevelopment of a neglected brownfield site
- positive impact on the local town centre
- employment and training opportunities.

Box 4.2 BREEAM Communities: best practice engagement

A number of councils are now advocating the Building Research Establishment Environmental Assessment Method (BREEAM) Communities requirements when developing masterplans for large-scale residential-led urban extensions. The scheme can be used to assess the sustainability impacts (social, environmental and economic) of a scheme and covers a range of disciplines, as well as setting out clear requirements around the community engagement process.

BREEAM Communities can be used to assess the constraints and opportunities relating to sustainability on the site and requires consideration of how the development will impact on the wider community. This requires the developer to draw up a consultation plan 'to ensure the needs, ideas and knowledge of the community are used to improve the quality of stakeholder engagement, throughout the design, planning and construction process'. Independently facilitated community consultation methods are advocated to engage the community, and it is necessary that '[g]ood practice consultation methods are used to engage members of the community and appropriate stakeholders in the process of designing development proposals'. A workshop to inform the development of the masterplan is also recommended (with an additional credit awarded for this activity).

Hampshire County Council, in its role as landowner, used BREEAM Communities to help develop proposals for Uplands Farm Estate,[13] a residential-led urban extension for up to 980 new homes, a new secondary school, a local centre, public open space, sports pitches and allotments in Eastleigh.

Construction communications

One of the most frequent issues raised during pre-application consultation will always be impacts on existing local residents during the construction period. A Construction Management Plan will be put in place as part of the planning conditions (setting out hours of work, noise and light restrictions, permitted routes and access points for construction traffic, etc.); however, the management of communication during the construction phase can have a big impact on the reputation of a scheme and, in turn, the housebuilder or developer bringing the site forward.

There can often be a significant time lag in the start of the construction process following an outline application being granted and detailed planning being secured – and from a community perspective, now the theory is becoming a reality. On larger schemes where it often takes longer to discharge pre-commencement conditions, there can be a perception locally that due to the time elapsed, the project will not be going ahead. In addition, council seats may also have changed hands and new residents may have moved into the area.

A proactive approach towards keeping communities and stakeholders updated and informed during the construction process is a key element of the overall long-term communications strategy, as is ensuring that surrounding communities are able to get in touch and that they will be quickly and efficiently responded to (by either the communications team, the contractor or the developer directly). Again, having agreed protocols in place to handle enquiries and ensuring that these are logged and dealt with is a simple yet critical part of managing the project communications during this phase. Recurring issues, such as dusty or muddy roads, can quickly become a high-profile local issue if not addressed (especially on social media feeds), potentially tarnishing the reputation of the scheme as well as the developer and contractors.

Activities such as road closures require statutory processes (including actions by the local council or Highways Authority), and it is important to ensure that key messages can be coordinated and managed. For example, ideally, information about roadworks or utilities works should be communicated directly to residents by the construction team, rather than residents picking up statutory advertising in newspapers.

A consideration within these protocols will be to brief the contractor's staff regarding commenting on the development. One approach to managing this is to provide contractors with business cards that provide details of the project website (with links to an FAQ and the Construction Management Plan) as well as a telephone number and email for all complaints and enquiries. This also ensures that any potential sales enquiries are effectively channelled from an early point. It can often be advisable to establish a dedicated helpline or project website so that all issues can be captured and managed through a single point of contact to ensure professionalism and consistency.

A key aim of communications activity during this period will be to ensure that there is local understanding of construction working hours, programme and methods, and that there is an online resource easily available for people to source further information. In addition, providing regular updates and communicating and addressing any issues or delays in a proactive and open way, while managing expectations, will help to mitigate potential communication issues as the project progresses.

It is important to ensure that the communications team has access to the full works programme. While construction programmes are invariably subject to change, it is important to be aware about what is coming up six months later to ensure that resident expectations are properly managed. For example, after a set of initial works is complete, if the construction company will be returning again to that locality after a temporary break, this needs to be communicated so that residents anticipate the disruption. In addition, engaging with community stakeholders such as local schools can help avoid unnecessary community aggravation, such as being aware of how the timing of road closures will affect the school run.

In addition, the construction phase offers an opportunity to promote the project at the same time as the marketing of the new homes, and provides an opportunity to reaffirm messages around quality, placemaking and the identity of the homes and future community. This can be demonstrated by creating engaging content around the delivery of the first homes, early landscaping, key infrastructure and community spaces, using tools such as images, video, sketches, infographics, blogs and written articles.

Box 4.3 Barratt Homes: swift bricks

Barratt Developments, the UK's largest housebuilder, targeted ten cities and towns in 2019 in which to install bird bricks into its homes. The aim was to boost the numbers of swifts across the country by giving the birds homes in locations where swift numbers and nesting sites have been in decline.

Swifts are an urban species of bird that use spaces in rooftops or in old buildings to make their nests. The species has seen this serious decline in numbers partly because of modern building methods that eliminate access to rooftops and the demolition of old buildings, reducing available nesting sites.

The ten cities and towns, identified by the RSPB as locations in which Barratt could make the biggest impact to bolster declining swift numbers, were Birmingham, Bournemouth, Brighton, Bristol, Cardiff, Edinburgh, Ipswich, Manchester, Newcastle and Oxford.

The bricks are fully drained, ventilated and unobtrusive, matching the colour of the bricks being used in the rest of the house. Barratt, the RSPB and Manthorpe Building Products jointly developed the innovative design in a brick format so that they can be fitted easily into new homes as they are built. In 2017 the bricks won a prestigious NextGeneration innovation award for their design.[14]

This initiative is part of the company's corporate partnership with the RSPB to support wildlife on new housing developments. Building new homes for the swifts is an important goal in this partnership, as the UK's population of swifts has fallen in 2019 to fewer than 90,000 pairs, down from 150,000 pairs two decades previously.

New homes

The final phase of the homebuilding PR cycle, once the show home and marketing suite have been launched, is to provide positive PR to help promote the new homes and the lifestyle on offer.

Early in the development process, the focus is much more around direct engagement with stakeholders and neighbouring communities to build trust and understanding through the planning and construction process. As the project progresses, the communication focus needs to shift more towards online and social media content and media relations, as the scheme moves into the marketing phase and the audience reach expands.

In addition, the approach to PR will depend on the company and product, as a very different marketing and PR approach will be needed for a large-scale development (over a longer period with a wider reach) compared with a smaller development that fits within an established community.

Working with the sales team to understand the target market will enable messaging to be tailored and communication channels to be maximised to build awareness of the homes and to promote the vision of the overall scheme. For example, will the homes appeal to families, professional couples, singles, downsizers, commuters or older people? Does the scheme need to be promoted locally, regionally or even internationally? Is this a scheme that will appeal primarily to homeowners or also to investors, landlords or second home owners?

This is also the point in the project to really focus on promoting the project brand (which should have begun during the construction phases), working alongside the creative and sales teams.

Most new home developments will be built out over a significant time period, so creating a long-term PR planner that identifies opportunities for maintaining the profile of a scheme beyond its initial launch is required to maintain a long-term sales pipeline. This also presents an opportunity to show the vision of the scheme coming to life, as well as providing opportunity through community or CSR initiatives to strengthen the brand of the scheme as well as recognition of the housebuilder.

Box 4.4 Alconbury Weald: bringing together the new community

Urban&Civic uses a master developer approach for its strategic sites to accelerate large-scale residential delivery. As it invests in landscaping, infrastructure and community facilities up front, the first new residents feel that they are moving into a burgeoning community rather than a building site.

Partnership and community involvement have been integral to the development of Alconbury Weald[15] from inception through to delivery. The project team has held numerous engagement and consultation events with stakeholders and the surrounding communities: for the very first consultation on this former MoD site, Urban&Civic put on an event and took people on tours of the formerly restricted site so they could understand the scale and opportunities for development.

The first primary school at Alconbury Weald was developed alongside the first new homes and opened back in 2016 with just one small class of mixed age groups. A community shop, play parks, a skate park and allotments were also brought forward at an early stage, and a community officer was employed to help residents get to know their new neighbours, help organise events and develop new activities and groups.

As part of this ongoing community involvement and development, Urban&Civic holds resident forum meetings, sends out regular newsletters and has also organised a number of events to bring the new community together, including pop-up restaurants and pubs, outdoor cinema events and a summer fete, at which it held the inaugural match on the new cricket pitch.

Box 4.5 Communications programme: key project milestones

- public launch of project
- public consultation events and feedback
- planning submissions and outcomes
- start of site/ground-breaking ceremony
- updates on key works – access, junction/highway improvements
- landscaping – tree planting, creating wildlife areas
- turf cutting/topping out on key buildings – e.g. school, community shop
- launch of show home/marketing suite
- release of first homes
- first residents move in
- key activities/events on site.

Communication tools and tactics

There are many different ways in which developers can approach PR and communications for new homes, but what is clear is that engaging early and listening to feedback will invariably improve the desirability of a new housing scheme as well as help maintain and promote reputation.[16] It is important to remember that there is no 'one size fits all', that every community and every site is different, and therefore each requires its own tailored communications programme.

There are a range of toolkits and methods heralded as best practice, but it is important to choose a strategic approach and to select the tools that will yield the best outcomes for the project.

There is a growing role for online and digital communication and engagement channels. However, particularly at an early stage in the life of a project, traditional grassroots engagement channels such as meetings, public events, media relations, advertising and direct mail remain essential to ensuring an effective overall communication campaign.

There are also a growing number of opportunities as well as challenges when it comes to managing online and social media platforms, particularly in relation to controversial developments or in the face of unhappy customers.

There follows a summary of some of the key tools and techniques that can be used to engage around homebuilding and examples of how they can be used effectively, as well as some of the challenges they present.

Political, community and stakeholder engagement

Understanding the local socio-economic, political and policy context of a new housing scheme is key to informing the overall communications strategy. This research work can provide an invaluable tool for engaging with communities at a grassroots level.

Effective, targeted engagement with local political representatives, established community groups and activists can help quickly establish levels of dialogue and trust within the community, which it is difficult to establish otherwise. Managing expectations and providing information to these individuals and groups enables dialogue to be more easily established through existing networks, events, forums and social networks – helping build credibility and trust and increasing participation in the engagement process.

In the longer term, this can help build a sense of community, forming a key element of the overall placemaking agenda, beyond simply building new homes.

Public events, workshops and forums

Despite online portals now offering an alternative option for residents to review proposals and engage around new housing schemes, there can be no replacement for opportunities for face-to-face discussions with a project team as part of awareness raising and the trust-building process (although this is a resource-intensive approach).

Public exhibitions that address identified key issues (such as traffic and green infrastructure), highlight site considerations and define the scope of consultation can be highly effective in building dialogue, obtaining insight from 'local experts' and providing meaningful feedback to enable project teams to make changes and improvements to projects.

The use of design workshops is another good way to identify and interrogate local issues and opportunities in more detail to deliver real benefits to local communities and future residents.

However, to ensure the success of public events, it is important that they are well organised and managed, that sufficient information is provided and team members properly briefed, and in the case of workshops and debates, that they are properly chaired. Clear information on the nature of the event, invitees, discussion topics and agenda should set the expectation from the start.

In the case of large-scale homebuilding schemes, the establishment of community or resident forums to meet through the construction phase and following the completion of the scheme can help foster a long-term sense of local ownership and community, as well as helping to integrate existing and new residents. In addition, a residents' forum has the potential to lead to a Community Development Fund as well as to operate or manage key community facilities on site. Identifying and nurturing grassroots interest and activity is key to ensuring the success of this.

Box 4.6 Gower (Cathays) Ltd: the power of grassroots engagement

Following the refusal of an application for 24 apartments, developer Gower (Cathays) Ltd went back to the drawing board to design a new scheme for the redevelopment of a former pub in a densely populated, residential area of central Cardiff.

To address key community issues raised during the planning process, a design workshop was organised for local residents and community stakeholders. This enabled meaningful discussion between the project team and local residents and stakeholders, including local ward members and the local Welsh Assembly Member. The local community was able to articulate its concerns and ideas for a development that would be acceptable to local people within the site constraints. These comments informed proposals for a completely revised scheme of ten terraced homes, which was much more positively received by the local community and was given planning consent.

This is an example of how effectively a well-facilitated community design workshop can enable a design that fits with the aspirations of the local community to be brought forward.

Printed materials and direct mailing: newsletters, flyers and letters

Sending printed letters or newsletters to local residents remains a staple element of community engagement campaigns for new housing plans, both to advertise engagement/consultation activities and to encourage participation and responses.

Used effectively, this can be a great way to make contact, to establish interest and to highlight key messages about a new scheme, as well as acting as early marketing material to start to promote the longer-term vision to potential buyers.

Generally, printed materials will point people to further information or feedback online and will encourage attendance at an event. The inclusion of a tear-off freepost postcard can also be a good way to encourage people to respond and to capture early interest from potential buyers.

Printed mailings are also a robust way to keep consultees (particularly those that have requested feedback) apprised of the project progress and construction programme, with the regularity and type of correspondence depending on the project scale and activity.

Newsletters and flyers can also be an effective tool to help facilitate the creation of a new community, with regular examples of developers using materials to inform new (as well as existing) residents about community activities and events taking place on site (including through newly delivered community facilities) as well as to feed back on key issues raised, for example through a residents' forum.

Online content

Hosting project information online from an early stage is arguably essential to an effective communications and PR strategy for all homebuilding projects (regardless of scale).

However, the online content – whether this is a project-specific website or a web page on a company website established during the early stages of the project (during the planning process) – is likely to look very different from the online content developed to help promote and sell the final product.

At the early stage, the focus will be around the wider vision (incorporating the brand of the housebuilder and wider placemaking aspirations) and highlighting a commitment to taking on board comments and evolving the detail of the scheme. The content will then evolve during the process towards being a more marketing-focused site promoting the end product and lifestyle offer.

Visuals and virtual reality

The use of computer-generated images (CGIs) and virtual reality flythroughs is a great way to help people envisage what the new development will look like. This is particularly important during the planning process. It is often very difficult for people to comprehend distance on plans and how landscaping can be used to mitigate visual impact. Initial 3D visualisations can be prepared at an early stage – for example, for public consultation purposes – and subsequently amended to include more detail, materials and changes to the design as the project develops. 3D visualisations and photo-montages now form an essential component of many planning applications and are often repurposed for PR and community engagement purposes.

This technology can also be a powerful sales and marketing tool for homebuilders and their agents. At a minimum, front, rear and street scenes views are typically used to illustrate sales brochures and websites, giving prospective homebuyers an accurate visual understanding of the various different house types that are available across the site. Increasingly, internal views and walkthrough animations are also forming part of the marketing collateral for new homes – particularly where aspects of the internal layout or design are key to the distinctiveness of a new home. With advances in the realism and reliability of CGI over recent years, some developers are using 3D flythroughs and virtual reality walkthroughs to start selling new homes off plan at an earlier stage, without waiting for a physical show home to be completed on site.

Box 4.7 Waterstone Homes at St Nicholas Fields: off-plan sales

Regional housebuilder Waterstone Homes is a keen proponent of the power of 3D visualisation at both the planning and marketing stages of residential development. In the Vale of Glamorgan, Waterstone Homes launched the marketing of its St Nicholas Fields development with 3D flythroughs, internal and external CGIs, and a virtual reality experience, and subsequently secured reservations for five out of the ten properties within weeks of the launch event. This was several months ahead of having a physical show home available on site, demonstrating how effective a virtual reality tour can be in giving prospective buyers the confidence and clarity they need in order to buy off plan.

Social media

Social media present a range or opportunities as well as issues when it comes to managing communications for new homes.

During the pre-application phase, it can be an onerous task to manage social media platforms such as Facebook that enable residents to comment on the scheme, as it will be necessary to monitor and ideally respond to issues and comments being made. Considerable thought should be given to how much response and engagement is given. If the developer responds on one point and it generates a further four comments – how will this be handled? One approach is to reply to social media comments by directing users to information on the project website, where the content cannot be manipulated and is there for all to see.

Social media can pose a reputation risk, as was demonstrated by head-lines around Persimmon reportedly paying to take over the administration of a group entitled 'Persimmon Homes Unhappy Customer', resulting in the headline 'Persimmon shuts out critics of its homes with Facebook take-over'.[17] A considered and ethical approach to managing social media therefore needs to be adopted at all times.

Based on the overarching communications strategy and objectives, it is important to identify which social media channels will be appropriate for the project, and the content that will need to be produced for these, and how the channel will be monitored and managed.

Facebook offers a great platform for targeting local communities, both during the planning phase as well as by sales and marketing teams once the scheme is more advanced, through targeted advertising.

Twitter can be a useful tool at the early stages of a project, as it is used by a wide variety of influencers, so it can be useful in promoting content to key identified influencers and stakeholders.

Instagram is used by a large number of new home developers as a tool to illustrate their brand and lifestyle, although it is unusual for project-specific Instagram feeds to be established as it can be challenging and resource intensive to maintain a regular stream of engaging content.

Once social media channels have been established, creating a forward PR planner and ensuring a regular feed of content is key to building credibility and interest to help ensure that they meet campaign objectives. To manage and protect against negative or inaccurate stories escalating (or going 'viral') on social media, structures will also need to be in place to ensure regular monitoring and management of responses.

Media and blogger relations

As mentioned throughout this chapter, the impact of media coverage on a housebuilder's brand and specific schemes can be critical to the reputation management of a project.

Identifying key media and building a relationship with these outlets will open up media opportunities to help promote the scheme and ensure an opportunity to respond in the face of opposition or adverse coverage.

During the sales stage, the target media are likely to broaden beyond local media outlets to reach a wide range of potential buyers. Media could include industry publications (housing, property and construction) as well as wider consumer lifestyle and home publications and supplements within the local, regional and national press (depending on the nature of the scheme). Blogger and online influencers are also an important media channel to consider, and an influencer relations strategy will be required to meet their needs and interests.

Undertaking preparation at each phase of the project is necessary to ensure that relevant information is available to respond appropriately to media interest. In addition, the use of strategic proactive media releases at key milestones and events to promote the scheme will help build the profile of the project over time.

Development and housebuilding is always a good headline grabber. At a local level, there may be a reporter who follows the project throughout planning and development. The focus of local press releases should be around the details of the scheme and the benefits for the local community, as well as providing a platform to promote events. It is always important to ensure that journalists are properly briefed and have the correct facts to hand.

The trade and national press will require very different information and are likely to be more interested in the project financials, key players involved and the overall state of the market.

A creative approach to media pack content can be a good way to build interest, so as well as including key information about the proposals, it can also feature quotes, images, company background, customer case studies and FAQs. In addition, engaging digital content, such as video and 3D visuals, and interactive virtual reality content can be a great way to build profile, particularly as news channels look for increasing levels of interesting online content.

Offering site visits to key media is a useful tactic to create media interest, as well as using events and milestones to create photo opportunities (such as turf cutting, topping out, landscaping schemes, show home openings and first residents moving in). Community, CSR, and customer events and initiatives are another good way to secure a diverse range of media coverage.

Sponsorship

Investing in the local community through sponsorship opportunities as part of an overall CSR approach can be a great way to encourage community spirit and local interest in the project, bringing about tangible community benefits. Although not often appropriate at early stages in the process (to avoid potential bribery accusations), this can provide a good hook for strong online content and media relations opportunities as the scheme progresses through construction to occupation.

Box 4.8 Barratt Homes: community fund

Barratt Homes and David Wilson Homes recently launched a community fund to strengthen its ties with the areas in which it is building new homes. The fund donates £1,000 each month to a charity or organisation that improves the quality of life for those living in the region.

In 2018, £1.2 million was donated across the company in the UK, and the 2019 goal was to raise that figure considerably.

For example, Barratt Homes and David Wilson Homes in Southampton supports a number of charities, including Kids Out, and in 2018 raised more than £64,000 for the charity, which provides disadvantaged children with great memories and days out. It also supports local schools with hi-vis clothing, the Countess Mountbatten Hospice and the Romsey Relay Race.

Managing data

Another consideration throughout all engagement activity around new homes is ensuring that GDPR guidelines are followed. As well as ensuring ongoing engagement with consultees, it is important to be able to follow up with supporters of the scheme, in addition to individuals who have expressed an interest in buying a home on the site. Full guidance on data protection and the handling of personal information is available on the Information Commissioner's Office website.[18]

The future of homebuilding communications

Housebuilding has been at the forefront of the national and political agenda for many years now, and the ongoing struggle to deliver sufficient housing in the UK suggests that it will retain a prominent position for many years to come.

As with all aspects of life, the industry will continue to face new challenges and will need to be able to respond to many issues, including the need for more environmentally sustainable homes; the need for lifetime homes; the decline of the high street and what that means for community cohesion; the impact of AI on the construction industry; the competence and quality agenda in construction; and the impact when new generations reach a home-owning age.

As shown throughout this chapter, it is necessary for any PR professional designing a campaign for a housebuilder to continue to reassess and pivot the focus and direction of the campaign throughout the life of the project, using an overall strategic approach, while responding to the local and national PR landscape.

This will continue to be the case, as many of the current issues continue (specifically the need for housing), but more issues increasingly come to the fore from a communications perspective, as is shown in Box 4.9.

Box 4.9 Housebuilding PR issues of the future

Decarbonisation: with the UK government committing to achieving net zero carbon emissions by 2050, and, for example, the Welsh government (as well as many local authorities) declaring a climate emergency, there is a growing focus on the decarbonisation of homes and new building standards that will boost environmental sustainability. Although new and innovative features can provide excellent PR value, they also present their own communications challenges. For example, as low carbon homes with new technology and smart energy systems start to be more widely introduced, there is clearly a need to explain in simple terms how the homes operate and to convince early adopters to move into the homes and embrace new (sometime unproven) systems.

Build to rent: currently a well-understood model across Europe, purpose-built 'build to rent' units are increasing in prevalence within the UK, offering a professionally managed alternative to the traditional private rental market and being a more affordable reality for many younger people today due to the nature of the housing market. See Chapter 9 for further information.

Ageing population: The population is ageing, and one in five of the total population will be over 65 in ten years' time.[19] In the UK, the vast majority of over-65s currently live in the mainstream housing market, meaning that increasingly, homes are going to be needed that can help support the health of older individuals. Current strategies include building new age-friendly homes, integrating housing with health and care, and developing new models for adapting and creating smart homes.

In addition, customer expectations will continue to change, so the messaging and tools used will need to adapt and improve to ensure effective communication campaigns moving forward, particularly in relation to selling the vision of a new project, but also in order to be able to capture the interest of potential customers. This will no doubt include increasing use of online engagement tools and digital communication channels, as well as more and better graphics and interactive digital tools (such as virtual reality models of schemes), to promote new homes and new neighbourhoods.

Until there is greater consistency in terms of quality, customer service and user experience of new homes, it is likely that reputational challenges in this sector will continue. However, this only underpins the critical importance of high-quality PR and communications guidance and the need for PR to be respected and adopted as a strategic management discipline that can make a positive difference at the planning, construction and sales stages and beyond.

Notes

1 https://hoa.org.uk/campaigns/ [Accessed 14 October 2019].

2 https://hoa.org.uk/campaigns/campaign-better-new-build [Accessed 15 September 2019].

3 www.theguardian.com/money/2018/nov/17/quality-build-homes-charles-church-buyers [Accessed 15 September 2019].

4 www.mirror.co.uk/news/uk-news/homeowners-229000-dream-home-turns-1240 6619 [Accessed 15 September 2019].

5 www.hbf.co.uk/news/overwhelming-majority-buyers-happy-their-new-build-home [Accessed 15 September 2019].

6 www.insidehousing.co.uk/news/news/welsh-housing-minister-slams-private-sector-for-building-slums-of-the-future-61268 [Accessed 15 September 2019].

7 www.thetimes.co.uk/article/the-grand-design-for-better-housing-3jv96wzfz [Accessed 15 September 2019].

8 www.independent.co.uk/news/uk/home-news/housing-homeless-crisis-homes-a8356646.html [Accessed 14 October 2019].

9 www.thetimes.co.uk/article/help-to-buy-scheme-has-helped-biggest-developers-to-double-profits-759qq3h7m [Accessed 15 September 2019].

10 www.thetimes.co.uk/article/jeff-fairburn-persimmon-chief-in-bonus-row-was-paid-85m-m0v5m9zlz [Accessed 15 September 2019].

11 www.bbc.co.uk/news/business-48597203 [Accessed 15 September 2019].
12 www.hbf.co.uk/documents/6879/HBF_SME_Report_2017_Web.pdf [Accessed 15 September 2019].
13 www.uplandsfarmestate.co.uk/ [Accessed 15 September 2019].
14 www.barrattdevelopments.co.uk/~/media/Files/B/Barratt-Developments/press-release/2017/Barratt%20wins%20NextGeneration%20innovation%20award%20for%20new%20swift%20brick.pdf [Accessed 14 October 2019].
15 www.alconbury-weald.co.uk [Accessed 15 September 2019].
16 For examples, visit www.consultationinstitute.org/; www.cipr.co.uk/content/policy-resources/toolkits-and-best-practice-guides; www.communityplanning.net/ [Accessed 15 September 2019].
17 www.thetimes.co.uk/article/persimmon-shuts-out-critics-of-its-homes-with-face book-takeover-sw6bkmn3b [Accessed 15 September 2019].
18 https://ico.org.uk/for-organisations/guide-to-data-protection/guide-to-the-general-data-protection-regulation-gdpr/ [Accessed 15 September 2019].
19 www.local.gov.uk/sites/default/files/documents/5.17%20-%20Housing%20our%20ageing%20population_07_0.pdf [Accessed 15 September 2019].

5 Promoting housing associations

Creating a community

Susan Fox

The social housing context

Housing associations are not-for-profit organisations that provide subsidised homes to people on low incomes. They provide a wide range of housing, including general rent, independent living schemes for older people, supported housing for those with specialist needs, and shared ownership properties.[1]

The term 'housing association' came into official use in 1935[2]; they are also known as Registered Providers (sometimes shortened to RPs), registered social landlords (RSLs) and social housing providers.

There are more than 1,600 Registered Providers in England.[3] The Ministry of Housing, Communities and Local Government estimates that about 10 per cent of households (2.4 million) rent from housing associations, a figure that has not changed in a decade.[4] Some housing associations date back to medieval alms houses; others, such as Peabody, have their origins in Victorian philanthropy.

Housing associations vary widely: they can be small, local bodies with fewer than 1,000 homes, or large organisations such as The Guinness Partnership, with over 65,000 homes across England.[6] They can focus on specific communities: for example, age, faith or ethnicity. Most are proud to be more than just a landlord, offering support services and community initiatives. Regulatory requirement for robust finances makes larger associations desirable, so mergers are common. And housing associations are big business: in England, the sector is estimated to hold £156.2 billion in assets, with a combined turnover of £20.5 billion.[7] Many are diversifying, acquiring commercial subsidiaries, which gift-aid profits to the social housing parent. There is a great network – the National Housing Federation, for example, runs groups for its members, including a communications network and conference. However, as housing associations become increasingly commercial, peers can become competitors.

Although housing associations are independent, they are regulated by the state, as much of their rent bill is met by housing benefit (claimed by 60

Box 5.1 Professional bodies in UK housing

Organisation name	Members (approx.)	Definition of housing association[5]
National Housing Federation (NHF) (England)	900	Housing associations exist to ensure that everyone can live in a quality home that they can afford.
Scottish Federation of Housing Associations (SFHA)	158	A housing association is a voluntary organisation dedicated to helping people obtain decent, affordable accommodation which meets their needs.
Community Housing Cymru (Wales)	70	Housing associations are not-for-profit companies set up to provide affordable homes for people in housing need.
Northern Ireland Federation of Housing Associations (NIfHA)	20	A housing association is an independent, not-for-profit social business that provides both homes and support for people in housing need, as well as key community services.
	1,148	

per cent of social renters[8]); they may also receive government loans to build new properties.

The regulators provide useful frameworks and standards, covering, for example, information on expected levels of tenant involvement and consultation. Complaints resolution is through the Housing Ombudsman service and the Local Government Ombudsman (England) and the Public Services Ombudsman (in Scotland, Wales and Northern Ireland). Their websites provide handy information, such as how to write a meaningful apology.

Box 5.2 Housing association regulators

Regulator of Social Housing (England)
Scottish Housing Regulator
Welsh Government Housing Regulation Team
Northern Ireland Department for Communities

Strategy

For some housing associations with a very narrow focus, a communications strategy may be relatively straightforward. But if you're in a housing association that has diversified (as many have over recent years), a clear brand strategy is vital: is there one housing association brand, or several? Do your subsidiaries use the housing association badge or separate brands to suit different target markets?

You may need a communications strategy for each part of the business (common elements include social housing; maintenance, repairs and construction; private rent, shared ownership and sales; training companies; homelessness and community charities; and call monitoring services).

What is the main focus for each part of the business – is it business to business (B2B) or business to consumer (B2C)? Does it require marketing, corporate social responsibility, PR or fundraising? How will you manage stakeholder relations and internal communications?

An understanding of each part of the business will help identify the communications skills needed and will inform resourcing decisions.

Just as in other organisations, the key to effective communications lies in clear objectives.

The starting point is the housing association's corporate plan. The presence of a regulator means that housing associations are generally good at planning and performance measures. Check what the business wants to achieve over the next few years and create your communications strategy to support those aims. Align plans with other departments, particularly sales teams; close working with HR, IT and Business Transformation colleagues

Figure 5.1 Strategic layers in housing communications strategy.

is essential. Be flexible – housing associations can be opportunity led, and a significant new development, an unexpected partnership opportunity or a merger can change everything overnight.

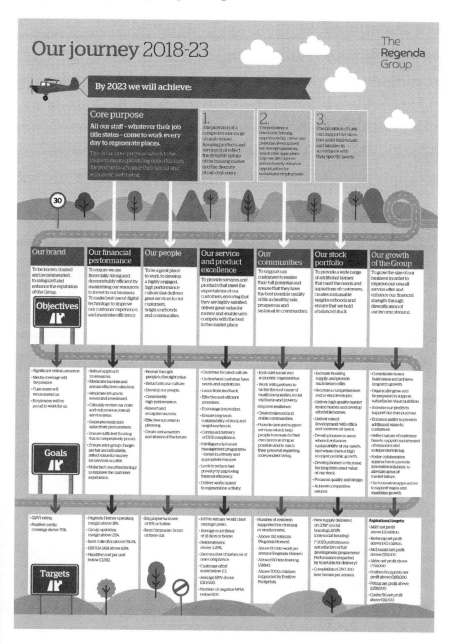

Figure 5.2 The Regenda Group: poster summarising the Corporate Plan and its supporting strategies.

Box 5.3 Redwing: developing an estate agency with a difference

Redwing has been established for more than 50 years. It offers private rented properties, leasehold and apartment block management, and shared ownership and sales, managing over 2,000 homes and retirement properties across the North West. The company also owns and manages commercial premises. It is a non-charitable housing association registered as a Co-operative and Community Benefit Society.

Redwing's leasehold and shared ownership sales process was simple and guidance led, and it won awards for helping customers through the home buying maze. The company therefore decided to roll out the model to all tenure types at all price points, developing its sales service into a fully-fledged estate agency, while emphasising its not-for-profit, community-aimed ethos. To ensure a quality product, the company started with staff skills, building a team that includes National Association of Estate Agents (NAEA) and Association of Residential Letting Agents (ARLA) qualified negotiators, Energy Performance Certificate (EPC) assessors and dedicated sales progression specialists.

The marketing strategy included ensuring that the pricing structure was fair – a flat fee, with no commission, with costs set out clearly from the start, allowing buyers and sellers to budget effectively. Products that would usually be extras were included or available in-house, reducing the pressure on negotiators to upsell for commission.

As many of the company's customers are first-time buyers and purchasing non-traditional tenures such as shared ownership, the company engaged external partners to provide guidance on mortgages, finances and conveyancing. Their core message was its community focus and not-for-profit status.

Redwing grew its engagement, using targeted digital marketing to communicate its values and its services. It launched a new website providing advice and guidance on the sales process and tenure types as well as traditional property listings, and developed a portal where customers can pay rent and charges, log repairs, view building information and liaise with property managers.

The new-style service won Redwing a contract with Liverpool City Council's ethical housing project, Foundations, managing the sales and marketing process, developing a brand identity and marketing, and coordinating launch events and show homes.

Informing the strategy

Housing is rarely out of the headlines, particularly since the tragic deaths of residents of Grenfell Tower in London in 2017, which called into scrutiny issues including health and safety, property management, resident liaison and crisis response.

Politically, housing is not always a priority portfolio; with eight housing ministers in as many years at the time of writing, political influencers have recently had their work cut out. In 2018, the Department of Communities and Local Government was renamed the Ministry of Housing, Communities and Local Government (MHCLG), giving the sector nominally more prominence.

A succession of government policies has buffeted social housing. Rent is set by government; an unexpected reduction in 2015 led to cuts, and the impact is still being felt. Universal Credit poses the risk of rent arrears – UK housing organisations have joined forces to lobby the government for changes to the system. Housing associations also worked together to explain the impact of the under-occupancy charge, known as the Bedroom Tax.

Partnership working pays off, as communicators from housing associations can coordinate messages locally to prevent competition for headlines. The rise of regional mayors also gives an opportunity for housing associations to work together, making the link between housing, health and employment.

Brexit has complicated financial forecasting, with funding, development, staffing levels and ability of tenants to pay rent affected.[9] In January 2019 the Information Commissioner called for the Freedom of Information Act to cover housing associations in England, Wales and Northern Ireland.[10] The introduction of freedom of information obligations would challenge the way many housing associations manage their messages.

As housing associations know the names and addresses of all their customers, you'd think targeting would be easy. However, some housing data systems are unwieldy, and the full benefits of big data, targeting and profiling are yet to be realised. The older age profile of some tenants means that channel shift can be slow. Social media is avidly adopted, with lively online communities, such as the weekly @SOchathour Twitter chat for shared ownership marketers. However, much social media activity is one-way, and social customer service is patchy. Halton Housing and Bromford are generally accepted to be leading the way with new technology. Bromford runs the innovative Bromford Lab (described as a place to think, break and rebuild ideas), with a blog and Twitter feed to share experiences.[11]

There are some research and benchmarking resources that give an element of comparability between housing associations and their activities. These don't usually go down to the detail of the communications department, which is often combined with other back office services. Popular services include HouseMark or similar benchmarking services, the *Sunday Times* Best Companies list, the STAR survey of tenants, the Net Promoter Score and the Institute of Customer Service.[12]

Given the context of uncertain policy, increasing commercialisation and technical change, the role of PR and communications in getting the organisation known, trusted and recommended is vital.

Messaging

Housing associations need to tell a compelling shared story – and the National Housing Federation (NHF) in England has created a campaign, Owning Our Future, to help them do just that. Results of a survey of politicians, media and think tanks[13] showed that the sector was bad at defining itself and that in the absence of a strong story from housing associations, other narratives could thrive, particularly about efficiency and the building of new homes. The survey showed that housing associations had no clear political allies, but it did confirm that familiarity breeds favourability. The NHF recognised that communicators could make a contribution to addressing these issues, and it set about training and supporting housing associations to tell their stories. Real-life stories, stakeholder relations training and some simple central resources such as infographics and template news releases have boosted the confidence of housing association communications.

The spectrum of activities that housing associations get involved in means that messages can be varied. A seasoned board member once told me her biggest challenge was to get housing associations to stick to housing. PR practitioners will, of course, find themselves promoting housing, but they are just as likely to promote litter-picks, wellbeing initiatives, back-to-work clubs and community fun days, as well as business-related messages such as staff changes, new acquisitions and award wins. Identifying the strategic contribution of the initiative will help with prioritisation.

As well as sales messages, a housing association is likely to want to emphasise its social heart, regeneration work and the care it takes of

One in ten people live in a housing association home

Figure 5.3 Housing statistics – examples of consistent messaging via NHF.

Does what you're doing support our calls to action?

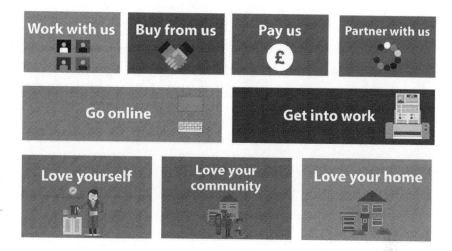

Figure 5.4 The Regenda Group: internal communications poster showing message priorities.

tenants. Housing associations that build will want to get the message across that they are developers – they will be looking for land, partnership deals and connections.

Social housing properties are usually advertised through centralised choice-based lettings services (used by housing associations in the area). Only hard-to-let properties are advertised by the housing association. This is often done by lettings teams, so close liaison between the communications team and the lettings team is important. Marketing guidance can help let properties more quickly, covering information to include in the advert, how to take a good photo, template texts and tips on which channels to use. Knowledge of the media and search habits of prospective tenants is valuable (the term 'council housing,' for example, is often used as a search term by prospective tenants, though rarely used by housing associations themselves). News of vacant homes and jobs performs well on Facebook.

Campaigns aimed at behaviour change are often required. For example, at Christmas housing associations usually remind tenants to prioritise their rent payments, offering an incentive such as high street vouchers. Many housing associations will promote skip days to prevent rubbish being left outside properties; and chip-pan amnesties are common, swapping the pans for deep-fat fryers to help prevent fires. In these cases, very local communication is required, with an emphasis on ease and popularity; leaflets, posters, community Facebook groups, word of mouth and the local paper help spread the word.

Some tenants say that there is a perception of stigma associated with living in social housing, which can be reinforced by tabloid newspapers, social media and so-called 'poverty porn' television programmes. Housing associations won't want to reinforce that image, so it is important to check case studies and the portrayal of tenants, as well as the tone of voice and language used. The Benefit to Society campaign has published a useful guide for journalists to help avoid stereotyping.[14]

Marketing shared ownership properties can be complex, as the product is not well known; explanation is often required about what it is, how it works and what the eligibility criteria are. In addition to the usual online property sales sites and development microsites, PR and media relations can add value, with items about shared ownership in targeted local media, providing plain language information on the website and using social media stories to attract attention. Most buyers are looking primarily at location rather than tenure, so it tends to be best to focus activities in the geographical areas where the housing association has shared ownership properties for sale. Case studies can be difficult to secure, as customers tend not to want to feature their private lives and finances. The National Housing Federation is working on a campaign to help debunk myths and make shared ownership a more familiar and attractive concept.

People

The scope of audiences to be communicated with will depend on the type of housing association. While tenants and employees are the most obvious groups, it is also common to consider the regulator; local, regional and national politicians; a wide variety of contractors; other housing associations and organisations working in the same communities; banks and lenders; and campaign groups, to name a few. These groups will be further broken down by type or profile. Some housing has specific eligibility criteria, so it is important to be clear about who is included and excluded. Shared ownership, for example, is open to those below a certain income, and favours older people and those who were in the armed forces.[15] Check out the criteria of the development you're marketing to be sure you're targeting the right people.

Social housing residents are individuals, and it would be a mistake to assume that one type of communication will suit them all. All human life is there: young parents, care leavers, single people, families, older people, people with disabilities, people with support needs, people with and without jobs, for example. Business requirements mean that audiences are often targeted by geography, type of tenure or property owner. Find out what level of profiling the housing association has managed. People often want to send a leaflet to every house on an estate or every tenant on a list, but a bit of profiling information can better target communications and avoid expense – targeting those out of work when promoting a job club, for instance, is cheaper and more effective than sending a mass mail-out to all tenants.

Given that a large proportion of tenants may be older or have specific needs, communicators need to be sensitive to the abilities and preferences of individuals. Clear, accessible communication is essential, and encouraged by the regulatory standards.[16] While social media can be effective, it may exclude some tenants, so more traditional means of communication, such as posters and newsletters, still play an important role. Neighbourhood officers and scheme managers know their residents well, so a good starting place is to listen to what they can tell you.

Most housing associations have scrutiny panels, reading panels or residents' groups, and these are a fantastic resource for the communications team – to get feedback on draft publications or to test out a new website, for example. This helps to ensure that the customer is at the heart of the service (a phrase that many housing associations like to use, and which requires effort to make it a reality). One tenant once told me: 'I feel like I've been community liaised at!' Effective communication is two-way – do it with your customers, not to them.

Another target audience may be other people's tenants. If you're working on a place-based community project, for example, you may need to reach anyone in the area who can impact on the project, which will include homeowners and other people's tenants. Working in partnership with community groups and other housing associations on the patch will help reach the right people.

Many housing associations are good at stakeholder relations. Knowing the local MP, councillors, key council officers and businesses is important. In large organisations, much of this is done at local level. Identifying stake-

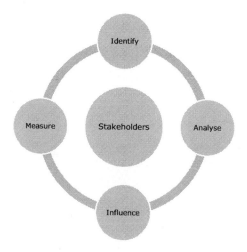

Figure 5.5 Typical cycle of stakeholder relationship management in housing.

holders is an essential part of the communications planning process; for example, when creating the communications for a development or a regeneration scheme. Foster good relationships by engaging with the right people on social media, invite stakeholder representatives to events, and keep key people informed with e-newsletters and publications.

With staff working in head offices and subsidiary companies in the neighbourhoods or on properties and developments, it can be hard to reach some employees. The communications team can help to develop a strong culture, reinforcing the common goals, mission and values – close work with the HR department and Executive Team is vital. Social housing staff tend to be compassionate and like to be with people; wellbeing initiatives and charity events go down well, as do opportunities to meet up socially. Increased use of technology such as tablets and mobile phones makes internal communications easier, with intranets, collaborative working tools and videos effective in getting messages across to those working remotely and giving them a voice.

Box 5.4 The Riverside Group, Hardwick House: using social media to reach armed forces veterans affected by homelessness

Hardwick House in Teesside provides accommodation and support for local armed forces veterans affected by homelessness. Initially, reaching homeless veterans proved to be a challenge. Riverside could have spent thousands of pounds doing this, but the communications team had to deliver results on a budget. Profiling and investigation with the local team showed that reaching veterans' local support networks, families and referral agencies would be key.

Their strategy was to use Twitter and Facebook adverts to raise awareness and to generate referrals, targeted at people within 25 km who were linked to local veterans' charities or who had a military-related occupation.

Facebook and Twitter posts directed users to a web page with a form to register expressions of interest. The teams also posted organically on Twitter, tagging local military charities. Supporting channels were also used to capture non-social media audiences. These included posters, local media, email campaigns and face-to-face presentations. Facebook Pixel and Google Tag Manager and Analytics were used to track acquisition and the user journey. Riverside tracked conversions by building questions into enquiry routes, demonstrating return on investment (ROI) by calculating referral numbers from each channel. Facebook and Twitter offered precise audience targeting, considerable audience reach, and value for money. The team regularly optimised and re-targeted adverts at the demographics who engaged the most.

Advertising visuals and copy were designed to engage emotionally with target audiences. An emotive image that the audience would empathise with was chosen, and video interviews of service staff were used to grab attention. Messages were locally focused, tapping into people's empathy and support for causes in their local area. Media coverage included a radio interview on BBC Tees local current affairs programme. This campaign captured the imagination:

- Facebook reached 30,103 people and gained 3K+ engagements and 423 shares
- Twitter reached 57,383 people and gained 1.8K engagements and 87 retweets
- The Hardwick House website page received 971 visits that month, compared with just 85 the previous month – a 1,042% increase.

Within six weeks, all the flats were occupied and a waiting list established. With a total advertising spend of £96, the final ROI was just £6 per referral. The legacy of this campaign has since transformed the way that Riverside markets these schemes and engages with other traditionally hard-to-reach audiences, including victims of domestic violence.

Tactics

Housing associations can incorporate a wide variety of businesses and target audiences, so it is standard practice to employ a breadth of tactics to get the job done. There is unlikely to be a huge budget for communications, so quick and cheap tactics play a great role alongside the bigger set pieces.

Brand design is the bedrock. Guidelines – shared with staff – will help. It's useful to have a clearly articulated brand essence, particularly where there are multiple brands; guidelines on tone of voice encourage consistency in written communications. Easy-to-use templates for posters and newsletters help neighbourhood staff to communicate proactively while keeping the 'look and feel' on track.

Websites are the first port of call. A good understanding of customers allows the creation of clear user journeys – tenants and stakeholders will have different information needs. Mobile-optimised sites are essential. Sites with transactions, such as rent payments or repair bookings, will need to be secure. Housing associations typically have several separate data systems, such as the main housing database of tenants and properties, so check which ones the website needs to be compatible with.

Websites should be accessible (in line with government guidelines[17]) and written in plain language; a translation function is useful. As leaflets and brochures are still appropriate, it's easy for information to come adrift, so it is wise to have a procedure to ensure that the website always has the latest version, and print versions are updated in line with the website copy at the next print run.

The sector has not yet fully adopted social media as a customer service mechanism. Typically, the mood is set at the top. Some housing associations use it to good effect, with Facebook being most readily adopted by residents. This can be an ideal channel to communicate with customers, as it's easy and can carry pictures (of repairs required, for instance). However, some housing associations are wary and have not enabled comments. Targeting user advertising on Facebook can be an effective part of the

Figure 5.6 The Regenda Group: Branding strategy.

marketing mix for promoting property for sale and rent, and Twitter works well in reaching other housing associations and stakeholders. LinkedIn demonstrates corporate social responsibility, culture and ethical credentials – the strong network among housing associations makes it a good tool for engagement. Housing associations can fall into the trap of giving social media to an apprentice to run, often resulting in short-lived accounts with no record of the password. A clear understanding of the strategic contribution of each channel is the first step, and roles need to be agreed – whether customer service or communications staff respond to messages, for instance, and who monitors the sites. Training staff in subsidiaries to manage their social media is desirable, but it can often be forgotten, even in sales teams. Keep an eye on subsidiaries' content and engagement to ensure it is brand aligned. Some guidelines on effective social media use are usually welcomed by staff, and it is useful to publish engagement rules for external users, making it clear which types of comments will be deleted, for example, and to manage expectations of response times.[18]

In terms of media relations, it's important to have good links with local newspapers, where they still exist, particularly in areas where the housing association has concentrations of housing. It is also important to get to know the online sites relevant to the area and to the sector. If the housing association is the main housing provider, it's likely to be asked about housing topics, so clarify the organisation's policy on commenting on political issues. It's also more likely to become a target when things go wrong. Comments pages on local news websites need monitoring, and the communications team will need to decide whether it's worth engaging with the negative posts.

The PR role increasingly means imaginative use of owned media; most housing associations produce resident magazines, in print and online, at least twice a year, in addition to stakeholder e-newsletters. Nationally, titles to note are Inside Housing, 24Housing and Social Housing; the NHF and the Chartered Institute of Housing (CIH) do good e-newsletters, and Housingnet News provides a useful daily news roundup.

Conferences and awards are prominent. They are time consuming and expensive, so work out which to prioritise. The CIH holds a huge national conference every year, with three days of speakers, an awards dinner, a trade fair and fringe events. The NHF holds regular events, including an annual communications conference, and the Housing Quality Network runs useful workshops. The #CommsHero conferences offer lively events aimed at housing communications practitioners.

Mitigating risk

Reputation is important. Public expectations of ethical organisations are high – the way housing associations are run must be beyond reproach. Increasingly, housing associations are funding their own construction programmes: for every £1 of public money, they invest £6 of their own or private finance,[19] and the regulator's rating determines how much banks are prepared to lend.

A great deal is expected of housing associations that would not be expected of private landlords. Associations can be held responsible for intricate details of tenants' lives, covering anything from their behaviours, to their employment, to their abilities to use a computer. In addition, the NIMBY[20] faction can be strong, with fears of anti-social behaviour added to the usual concerns about the environment, access and architecture of new developments. A good reputation as a caring and responsible landlord is essential to see you through inevitable problem moments.

The Regulator of Social Housing in England publishes a Sector Risk Profile. It lists health and safety, fraud, weak procurement and poor probity as potential risks to reputation, and warns of increased scrutiny from tenants, lenders, investors, central and local government, and the media.[21]

Housing associations aren't public bodies (unless they wholly manage ex-council stock, in which case they are Arms-Length Management Organisations or ALMOs[22]). However, they do receive public money, so it is prudent for them to behave as much like a public body as is practical and commercially sensible. This includes avoiding taking party political sides. In their eagerness to foster good relationships, housing associations can sometimes inadvertently endorse a political party during an election, so it's a good idea to ensure that the organisation observes government guidance during election campaigns.[23]

Customers want to know that their rent is put to good use. It's useful to express expenditure in terms of weekly rent – would you be happy to defend that to your tenants or in the media? Stories demonstrating the benefit to residents are appreciated (job creation, community initiatives or housing improvements, for example). However, be wary of glitzy staff conferences or tweets from luxury locations, which may understandably generate negative responses.

The customer services team will often be the first to know what's going on. Neighbourhood officers and development and asset management teams are important. Remember repairs operatives too, who are most likely to meet tenants face to face; they also drive highly visible branded vans, which are likely to appear in photos and media footage. Housing professionals are normally good at handling crises – human tragedy often plays out in the home – so support them with a crisis communications plan and media training. Every housing association has an ongoing problem, a difficult development or a disgruntled tenant. These can run on for years, so get to grips with the facts, ensure messages are consistent and know the journalists on the patch.

New developments and regeneration schemes can be controversial. Some housing associations have valuable inner-city properties that they want to convert to private rented stock in order to fund new social housing in less pricey areas, giving rise to accusations of gentrification. Schemes most likely to attract negative attention are those that require moving tenants out of their homes. Housing professionals are used to moving tenants (or 'decanting' them, in housing jargon); the communications team can help with consultation planning, stakeholder identification and message management to reduce confusion and keep tenants informed.

Creative content

A lot of time is spent firefighting and getting large amounts of information into formats suitable for tenants and key stakeholders, leaving little time for creative work. Housing associations are keen to please multiple partner organisations and can be unwilling to take communications risks or stand out from the crowd. Furthermore, funding streams may impose rules on how initiatives are promoted, and associations using construction contractors will need to negotiate over the creative treatment of new developments. Nevertheless, housing association staff can be highly creative, running excellent neighbourhood and community initiatives. Get to know the neighbourhood officers to keep up with the news – many see these activities as part of the day job and don't always recognise the media and social media potential of what they're doing.

Increasingly good use is being made of photography and videos to showcase properties, with some associations also using virtual reality. Communications teams can exploit the interesting angles of unusual conversions – churches, court houses and pubs, for example. Social media is lively, with

Figure 5.7 The Comms Hero initiative is interactive and has a high profile on social media.

tenant events, local community activities and corporate news all doing well, particularly those featuring photos and videos. Real-life stories of tenants are popular, and while they can take a lot of time to gather, they are effective in explaining the difference a housing association can make. The media is always on the lookout for compelling personal stories, and most housing associations publish them on their websites.

The large number of activities in an average housing association, not to mention frequent significant projects, rebrands and mergers, can make for a vast communications workload. This requires quick, effective solutions: off-the-shelf guidance, simple templates, and a confident approach to social and traditional media all work in your favour. Financial cuts can also drive creativity. Annual Reports are now rarely costly printed documents, as housing associations explore cheaper formats, such as infographics, videos or spreads in the tenant magazine.

Marketing supplier Resource runs the @CommsHero Twitter account and frequently champions creativity among housing association communications teams, awarding #CommsHero badges to those who stand out.

Box 5.5 Stand Agency and Anchor Hanover: a successful awareness campaign

The care home and retirement housing market is competitive, and there is a lot of unfounded stigma and misconceptions surrounding the sector, occasionally resulting in negative media coverage. Stand Agency was briefed to develop a campaign to differentiate Anchor, England's largest not-for-profit provider of housing and care for older people, from its competitors.

Anchor had identified lack of seating as a key factor in deterring older people from getting out, increasing loneliness and isolation. So, Anchor and Stand developed the campaign SU4SD – Standing Up 4 Sitting Down – focusing on the provision of free seating, aimed at older people and their relatives, retailers, MPs and other potentially supportive organisations. Their strategy has three stages:

1. Awareness – most people didn't understand how important seating was for older people.
2. Understanding – communicating to retailers and politicians why it mattered and what needed to be done.
3. Action – achieving change through influencing decision-makers and influencers (including the public).

The objectives were to:

*   Create awareness of a little-understood issue facing older people in the UK and demonstrate the value of older people by highlighting the economic opportunity for the high street (target: reach of 20,000,000 and social media reach of 500,000).
*   Encourage politicians and retailers to take notice and action (target: support from five MPs, ten partners and one national retailer).

- Build awareness of Anchor as a provider of care and housing to older people (target: increase brand awareness by 1 per cent; double enquiries received during campaign period).

The focus was on traditional, online and social media (Facebook and Twitter), Anchor's own channels and those of partners and MPs. The creative inspiration for the campaign came from older people themselves and Anchor's positioning as their champions. Research showed two strong angles emerging: the importance of the Grey Pound and the knock-on effect on loneliness amongst older people. An independent survey of people aged 70+ found that more than two-thirds thought seating in shopping areas was declining, 78 per cent thought their town was not suited to their needs, and one in five was going out to shop less due to the lack of seating. Stand created a panel of experts who made recommendations for change. The International Longevity Centre investigated the economic impact, finding that it cost shops and businesses up to £3.8 billion each year. Stand created a compelling case study – 93-year-old Nona became the face of the campaign. Partners backed the call for change.

Campaign preparation took two months, and implementation was in four stages over one month:

1. Pre-briefings of key media contacts and MPs to create social media activity, which built momentum with traditional media too.
2. Warm up of targeted contacts in the fortnight before the launch.
3. Launch – sell-in, supported with targeted social activity.
4. Over the next 48 hours, every opportunity was fulfilled, whether national or local.

BBC Breakfast and BBC News launched the story, with live interviews and pre-recorded packages with Anchor's chief executive shown 18 times in one day, plus national and regional press coverage. The outcome was that almost a quarter of UK adults aged 45+ had heard of the campaign and a third knew its name. There were 249 pieces of coverage, with reach of over 209 million, and a social media reach of two million, with engagement 25 times higher than in the previous month. High street retailers committed support, as did 16 cross-party politicians and 20 organisations. All coverage identified Anchor as the campaign instigator and described its role; there was a 20 per cent increase in Google searches for Anchor care, and newsletter registrations increased 350 per cent. Prompted brand awareness rose by 2 per cent, and new enquiries rose by 550 per cent. The cost per opportunity to see/hear the coverage was £0.0002.

Phase two of the campaign saw a forecast of losses to the high street of up to £4.5 billion by 2030. Anchor designed a Seating Calculator to work out the number of seats a town or city should have to meet the needs of older shoppers. This phase achieved 111 pieces of media coverage and 3.7 million social media impressions.

In 2018, Anchor merged with Hanover Housing to become Anchor Hanover. Phase three of the campaign communicated the organisation's unwavering focus on championing later life, highlighting seating on public transport, generating eight pieces of national coverage and 17 broadcast interviews, plus support from public transport providers.

Measurement

Success needs to be measured against your objectives. On a corporate level, the main measure of reputation is the regulator's grading, with (in England) a governance and financial viability rating of G1V1 being the prize.[24]

Corporate communications measures vary according to the budget for primary research and the board's requirement. The proportions of positive and negative media coverage, and engagement on social media, are good indicators of when communication is going right (or wrong) and are suitable to report as a figure in a board paper. Resident satisfaction with specific channels – such as the website or the tenant magazine – is a useful measure, as is staff satisfaction with the intranet. Online reviews and customer complaints are good sources of information: housing associations pay a lot of attention to improving their complaint handling, yet online reviews are still to become a mainstream measurement, given the lack of real choice for many social housing tenants.

Communications can contribute to operational targets, such as void turnaround time (the time it takes to let a property): improved photos, positive descriptions and website key words can all reduce the time a property spends empty.

When marketing property for sale or private rent, the key measure is usually cost per unit. Communications can also reduce the cost of business-as-usual activity, such as checking that the gas supply in properties is safe: tenants often don't keep appointments, and running an awareness campaign could help to reduce this problem. Before and after counts will demonstrate the effect of such campaigns and communications improvements.

Resources

There is no single model for a housing association communications team. Resource often depends on how much budget has been allocated to PR and communications in the past. Some associations have a centralised communications team, some have communications people in each part of the business, and some will de-professionalise the service, giving frontline or neighbourhood staff some communications responsibility. While there is usually a separate sales team covering commercial, shared ownership and private rent, marketing and communications activities are normally combined. Some housing associations will have a PR or communications director, but communications teams are more often part of a combined resources function along with IT, HR and finance. Internal communications can find itself part of HR, and, depending on how far the association has adopted social media, the customer services team may play a significant role in running the main social media accounts.

The more the housing association has diversified, the broader the skills and backgrounds the communications team needs. You may be expected to be all things to all people: brand expert, marketing guru, stakeholder relations diva, media relations/PR/crisis communications pro, copywriter, ad buyer, customer communications leader, fundraiser, staff engagement specialist, public consultation planner, event organiser, campaigner, researcher, designer, website wizard and social media star. A combination of account management and technical specialism will be required. Communications teams will often find themselves in the role of trainers, teaching colleagues how to use social media, how to update websites and intranets, how to write plainly and effectively and how to handle the media in a crisis, for example. Many associations support training, and many have Best Companies accreditation.

Some housing associations will have an in-house designer, and more will outsource this service. Photography and video are increasingly done in-house (not just by the communications team). However, while in-house photos are considered suitable for the hard-to-let social housing properties, professional CGIs, photographs or videos, and microsites are required for new developments and property for sale and private rent. Large projects, such as campaigns for significant developments or partnership initiatives, are often outsourced; this helps to cope with peaks of work and enables the association to purchase specialist skills or experience. Some housing associations, such as The Sovini Group, offer communications as a shared service to other organisations. It is common for the same suppliers to pop up again and again as housing associations recommend services to each other.

The drive to reduce back office costs has in many cases hit the communications team disproportionately – PR practitioners may find that, despite their exemplary all-round skill set, they may not have the resource or level of influence they would like. The cost of communications as a percentage of turnover is a figure to track. Communications budgets can be spread across a variety of departments, projects and developments, so the communications team needs to analyse expenditure across the business to ensure that costs are properly understood and built into return on investment calculations.

Conclusion

Social housing is a fascinating sector for communicators. It is not straightforward property PR. It covers people's homes, lives and communities, often with an offering that is pretty much from cradle to grave. The emphasis tends to be on community, regeneration and place rather than just the properties themselves. There can be a struggle with external issues, such as political change and a lack of understanding of what housing associations are, and internal issues such as resources, prioritisation and increasing commercialisation. But the sector is compassionate, providing a much-needed service that is rarely out of the news.

Notes

1 Where the resident owns a proportion of their home and pays rent to the housing association on the remaining portion.

2 Malpass, Peter. *Housing Associations and Housing Policy, a Historical Perspective* (2000), London: Macmillan.

3 Regulator of Social Housing. *List of Registered Providers* (18 April 2019). www.gov.uk/government/publications/current-registered-providers-of-social-housing [Accessed 15 September 2019].

4 Ministry of Housing, Communities & Local Government (2019) *English Housing Survey Headline Report, 2017/18*. London: Ministry of Housing, Communities and Local Government.

5 From www.housing.org.uk/about-us/about-our-members/about-housing-associations; www.sfha.co.uk/faqs; https://chcymru.org.uk/en/about-us/frequently-asked-questions; www.nifha.org/about/faqs [Accessed 17 May 2019].

6 www.guinnesspartnership.com/about-us/what-we-do [Accessed 17 May 2019].

7 Regulator of Social Housing. *Global Accounts of Private Registered Providers* (2018), Leeds: Regulator of Social Housing.

8 Ministry of Housing, Communities & Local Government. *English Housing Survey Headline Report, 2017/18* (2019), London: Ministry of Housing, Communities and Local Government.

9 National Housing Federation. *Brexit – Issues for Housing Teams to Consider* (16 October 2018), London: National Housing Federation (www.housing.org.uk/latest-updates/brexit-issues-for-finance-teams-to-consider [Accessed 15 September 2019]).

10 Denham, Elizabeth. *Data, Transparency and Trust: How Information Rights Can Promote a Culture of Accountability* (2019), Wilmslow: ICO. https://ico.org.uk/about-the-ico/news-and-events/news-and-blogs/2019/01/data-transparency-and-trust-how-information-rights-can-promote-a-culture-of-accountability [Accessed 15 September 2019].

11 www.bromfordlab.com [Accessed 15 September 2019].

12 *Sunday Times* Best Companies to work for, www.b.co.uk; Housemark and STAR survey, www.housemark.co.uk/subscriber-tools/benchmarking/survey-of-tenants-and-residents; Institute of Customer Service, www.instituteofcustomerservice.com [Accessed 15 September 2019].

13 www.housing.org.uk/get-involved/promoting-our-sector/owning-our-future/current-perceptions, YouGov plc. Total sample size 1,516 adults. Fieldwork undertaken 27–29 January 2016.

14 Benefit to Society. *Guide to Reporting Social Housing* (2018), See the Person. http://benefittosociety.co.uk/the-campaign [Accessed 15 September 2019].

15 www.helptobuy.gov.uk/shared-ownership [Accessed 15 September 2019].

16 Homes and Communities Agency. *Tenant Involvement and Empowerment Standard* (2017), London: Homes and Communities Agency.

17 www.gov.uk/service-manual/helping-people-to-use-your-service/understanding-wcag and www.w3.org/TR/WCAG21 [Accessed 15 September 2019].

18 For example, www.regenda.org.uk/contacting-us-on-social-media [Accessed 15 September 2019].

19 National Housing Federation. *How public money is spent on housing*, London: National Housing Federation. www.housing.org.uk/how-public-money-is-spent-on-housing [Accessed 15 September 2019].

20 NIMBY stands for Not In My Backyard, a term used for residents who oppose a proposed development in their area.

21 Regulator of Social Housing. *Sector Risk Profile 2018* (2018), Leeds: Regulator of Social Housing.

22 www.almos.org.uk/almos [Accessed 15 September 2019].
23 www.gov.uk/government/publications/election-guidance-for-civil-servants [Accessed 15 September 2019].
24 Regulator of Social Housing. *Regulating the Standards* (2019), Leeds: Regulator of Social Housing.

6 Promoting local authorities as developers

From functional to aspirational

Fiona Lund

Any effective PR campaign, short or long term, needs to be founded on a strategy. This strategy must be informed by a deep understanding of market drivers, including the history of a sector, its complexities and where it's heading. The world of housing provided by local authorities has changed significantly, is complex and is charting new waters, and so PR and communications for housing must evolve accordingly. The first section of this chapter will look at those changing influences.

Setting the scene: a brief history of local authority housing

Local authorities have been required by law to provide council housing since the 1919 Addison Act, a response to Prime Minister Lloyd George's Homes Fit for Heroes campaign, which was established to look after returning First World War veterans. But it was only after World War II that the UK saw the true birth of the council house, when Clement Atlee's Labour government built more than a million homes, over three-quarters of which were council houses.

In the following decades, local authorities continued to build social housing, but more recently the landscape has changed dramatically. At the time of Margaret Thatcher's 'Right to Buy' initiative in the 1980s, local authorities were responsible for building more than 40 per cent of new homes, but by 2017, this had dropped to fewer than 2 per cent, according to the Department for Communities and Local Government (DCLG).

In terms of communications and PR in this sector over the decades, it's no surprise that there wasn't really much call for this professional expertise. The legislative nature of the sector left little requirement to promote and persuade. Even during the flood of funding into social housing under the Decent Homes programme,[1] when local authority tenants were offered new kitchens, external wall insulation, bathrooms, doors and windows, the only real PR in action was that of respective governments and local authorities telling the country how much they were investing into the public housing stock. Communication was to a captive audience and attracted little controversy.

But nothing stands still, and in the last decade the landscape has changed once again. Decent Homes funding introduced in 1997 began to slow down from 2005, with many of the large, publicly funded improvement programmes drawing to a close. This fact, coupled with reduced central government grants and the rent decreases introduced in the Welfare Reform and Work Act 2016, meant that local authorities started to feel the pinch and realised that they had to be more commercial as one way of addressing the gap. Councils had to start to look at alternative measures for releasing equity to fund builds, which led to them creating private for-sale housing to cross-subsidise the building of new socially rented houses.

2002 saw the rise of many local authorities establishing Arm's Length Management Organisations (ALMOs), not-for-profit organisations, wholly owned by local authorities and used to manage housing stock. We started to see social properties being replaced with private housing. And then followed the Localism Act 2011, which aimed to put control of housebuilding back in the hands of communities, so forward-thinking local authorities resumed development power and began to establish their own private or wholly owned subsidiary development companies. According to recent data,[2] more than a quarter (98) of the 326 councils in England have set up new housebuilding companies since 2012, a trend that, combined with joint ventures like the IKEA example mentioned later, continues its upward trajectory.

The upside for local authorities as developers is that they can take the holistic and longer-term view that private, shareholder-profit-driven housebuilders do not. Indeed, private companies and PLCs are often accused of only delivering the statutory minimum in terms of build quality and Section 106 requirements, driven by the bottom line rather than objectives aligned with society at large. From personal experience of working with manufacturers of security and energy efficiency products, we have seen how the specification of new homes is often superior within social housing compared with private sale properties from private sector developers. In contrast to private companies, local authorities' private development arms can use Compulsory Purchase Order (CPO) powers where necessary and, of course, can borrow more cheaply than housebuilders. Councils can benefit from the financial growth of schemes, while private developers, who don't traditionally tie up precious resources in land for the long term, are in this context motivated by similar objectives – as shown in the Worthing example that follows.

This brief history is important because it shines a light on what has come before and, for PR professionals, gives us background, context and a deeper understanding of where we need to strategically target our communications now and in the future.

The main point is that as UK housing is facing its biggest shortfall on record, currently estimated to be a backlog of four million homes,[3] the landscape of local authority development is changing, and changing fast.

And along with this shifting landscape comes a huge step-change in the way local authorities as private developers must communicate with existing and emerging stakeholders in a bid to help alleviate the crisis.

Flat packed homes

At the time of writing, Worthing Council has announced that it is considering a deal with BoKlok, owned by IKEA and construction company Skanska, to build up to 162 homes.[4] Requiring processes significantly more advanced than erecting a Billy bookcase, these off-site manufactured homes (which, of course, include an IKEA kitchen) can be manufactured, dispatched, delivered and erected in previously unheard-of build times. The way in which the deal is structured is also a new concept. Under the proposal, the model means that Worthing would get 30 per cent of the homes to be used for social housing in areas where there is a shortage of affordable properties and high house price inflation. The rest of the homes built would remain in the ownership of the developer, but instead of selling the land to the developer, the council would receive its share of the homes, lease the land for 125 years and receive ground rent, which should be about 4 per cent on the value of the land. Worthing says that this proposal produces 45 homes for its use, rather than 13 if a conventional model was used.

Looking at this emerging market from a PR perspective, two salient points come to mind. First, this model is a clear sign of the mushrooming melting pot we can expect between private and public sector development. As communications advisors to local authorities, we therefore need to reconsider our stakeholders when it comes to thinking about local authority housing and realise that it's not just the tenants and councillors we need to satisfy. The blend of affordable and private homes on new developments is only set to grow.

In many ways, local authorities are now finding themselves in changeable situations – either as partners with private developers or even in competition with them. From a PR and communications perspective, this pepper-potting approach means that it is no longer appropriate to promote developments as either 'social housing' or 'private developments,' which presents both challenge and opportunity.

The second point is that we also need to be aware that we're talking about people's homes. Whether affordable or luxurious, we all aspire to having safe, happy homes. Unlike almost any other product we experience as consumers, there will always be a softer, intangible connection and emotional attachment to the places in which we live. While the product itself may increasingly become linked to a story around Modern Methods of Construction and technical or contractual innovation, the PR surrounding it must not lose sight of the human story too.

Learning a foreign language

The aim of the public sector is to serve a whole population in the most efficient way possible, while private sector enterprises are primarily established with growth and profit motives. Therefore, it's no surprise that the language used to communicate by each sector is also different. For local authorities, the tone has traditionally been more legalistic and administrative, while in the private retail housing market, it's direct marketing, call-to-action focused and always aspirational.

A quick look over some housebuilder and developer websites shows dreamy locations, a hand-holding service, blank canvasses, 'living your dreams,' and wish-list fixtures and fittings. Enter the linguistic world of local authority developers and housing associations, and the language usually becomes more functional. Traditionally, it hasn't been about selling dreams. It has been about remits, KPIs, safeguarding, representation, community investment, performance scrutiny and monitoring, and tenant satisfaction levels. There's even debate about what we should call the people who live in social housing – clients? residents? tenants? And let's not forget all the acronyms: KPIs, RSLs, ALMOS, PFIs, CPPs, WoS, HAs and LAs, to name a few. PR and communications professionals working in the public sector need to be entirely familiar with the terminology of the sector and need to strike the right tone required by the communications strategy.

Communicating in an increasingly diverse sector

One of the most powerful lessons I've learned during three decades of working in public relations is never to forget that we do not cease to be consumers as soon as we go into the office, give a sales presentation or attend council meetings. The psychological pull of each of our personalities remains in different contexts, which is one of the reasons I've always challenged the status quo of communicating within a supply chain by setting firm B2B or B2C criteria. True, as set out earlier, we need to use the right language for the right stakeholders, using the right channels, but we should also never forget that we're ultimately talking to groups of individual human beings. This is never more relevant than when we're talking about people's homes.

With this underlying recognition that we need to humanise any PR activity we carry out within housing, different groups need different messages at headline level. So, during the strategic phase of any communications to this sector, the local authority's private development team needs to give careful consideration to stakeholders and messaging, as shown in Boxes 6.1 and 6.2.

Each of these stakeholders will listen to, watch, read and be influenced via different channels, so these channels must be researched and defined. As an agency of 13 years, we've seen and embraced the ascendance of digital

---

Box 6.1 Local authorities' stakeholders

Who are the stakeholders? Consider:

- MPs
- central government
- charities
- councillors
- current and prospective homeowners/leaseholders/tenants
- designers and architects
- DIY stores/retailers
- government agencies – e.g. the Environment Agency, Tenant Services Authority, Planning Inspectorate
- manufacturing supply chain
- other housebuilders and developers
- planners
- pressure groups
- residents
- retail entrants (such as IKEA)
- local community groups
- council tax payers.

Box 6.2 Examples of areas to address in messaging

Here are some examples of key issues and themes for communication:

- the local authority's plans and strategy to address local housing need;
- the benefits of the public sector developer;
- the benefits of new partnerships, including innovative new funding models for new housing;
- design standards and local architectural heritage;
- methods of construction – including the benefits of Modern Methods of Construction (MMC), where appropriate;
- environmental sustainability and measures to address climate change (particularly among local authorities that have declared a climate emergency);
- accessibility and life safety;
- strategic use of Section 106 and CIL;
- working with the local construction supply chain;
- community engagement and involvement.

---

and social media, but we've also seen how print is still important for communicating with social housing tenants, especially as many are older people who have not joined, and will never join, the digital revolution. Drop cards, resident newsletters, and updates about new developments and refurbishments are still often distributed through the letterbox. Print runs of tens of thousands are still popular within local authority and housing association marketing communications.

Social housing and social media make, in some respect, unhappy bed-fellows, especially when the stories we want to push involve change – which, inevitably, they will when we're talking about new homes being built. The proliferation of digital media, which can offer instant gratification, means that bad reviews or negative comments can be retweeted by millions within minutes. While digital channels can enable organisations to engage in a much warmer relationship with their customer, we also know they're the best channels to use to complain in a very public way. So yes, absolutely, we must use social media to interact proactively and inform, but never forget that keyboard warriors take to their Facebook accounts quickly. So, it's important that in addition to the planned messages you're scheduling across the channels (Facebook, Twitter, LinkedIn and Instagram are still the most popular within housing) you monitor carefully and respond fast.

When it comes to placing a news story about social housing, take-up has traditionally been easier as compared with promoting a commercial product. Unencumbered by the need to sell a product or service, there is less cynicism surrounding news releases and case studies concerning the social housing sector. However, local authorities and their PR departments will have to get more hard-nosed, less complacent and less comfortable now that taking on the role of private developer requires them to adopt a more commercial stance. Likewise, the approach to social media is taking on a more consumer-focused approach.

The 'silver tsunami'

One pressing issue that is hurtling up the agenda when it comes to new housing, and especially building new local authority and housing associ-ation homes, is the challenge of a fast-ageing population. More so than other housing shortages that have been under the spotlight previously (key worker homes and homes for single parents and young families), an ageing population brings with it very specific demands, particularly surrounding health challenges. The most profound of these is dementia, forecast to affect two million people in the UK by 2050.[5]

By 2030, one in five people in the UK (21.8 per cent) will be aged 65 or over,[6] so it's no surprise that debate is now focusing on how we can both build new homes and adapt existing stock to be fit for purpose for this fast-growing demographic. The challenge is not just a personal one affecting those individuals in retirement; it's also becoming a major theme in housing policy debate, encompassing housing, care and health sector partners. New build private and public developers need to work together to devise solu-tions to counter the obvious current shortfall in suitable housing.

The challenge for PR, therefore, is to ensure that messaging reflects the needs and interests of this older demographic. Many properties are sold, rented or leased to young families because of proximity to good schools, but older people have different priorities. They need easy access to local shops, healthcare centres and leisure facilities. Research by the NHBC Foundation[7] shows that they're happy to move for the right reasons, top of which include ease of maintenance, control over long-term running costs, lifetime warranties and future-proofing in terms of health and safety resources.

Interestingly, research by the NHBC Foundation also found that, compared with younger buyers, the priorities of over-55s play uniquely to the qualities of a new build home: they are 20 per cent more likely than younger buyers to want to buy or lease a new build home. This is, therefore, a huge market for local authorities, enabling them to maximise their commitment to supplying new build homes. As an example of an organisation spanning the private and public sectors and making a great job of communicating with this sector, take a look at L&Q or Anchor.

Another example is McCarthy & Stone. The specialist retirement builder for private ownership and leaseholders has a well-established PR and marketing engine: a templated model that they roll out before and during any development they undertake. Having determined the location, they generally set up a 'pop up' in the local town. They rent and brand an outlet such as an empty shop for approximately six months and provide an 'advisory service' offering information and support to local people interested in the development (it's an estate agency by any other name). The PR engine kicks in with local and regional press events, interviews, sponsorship and case studies of happy 'moving home' stories, alleviating the fear factor, and thus growing brand awareness and trust.

As with any PR campaign, the first step is to understand the drivers of this target group and especially the motivations for older people to remain in their own homes or to sell up and move to specialist communities. It's also important for local authorities to promote, both nationally and to local communities and policy-makers, the cost implications of people not moving early enough.

I believe that PR to persuade people to downsize, to release much-needed larger homes and to move in earlier should be a key thrust of any local authority developer programme. To achieve this, we need to engage early and address concerns and worries people have at this very sensitive point of their lives.

Again, we must not forget that we're dealing with people's homes. Messaging should surround the fact that these new homes are intelligently designed and future-proofed for complete peace of mind and flexibility for the future; show how people can customise and adapt homes in the future to meet changing health needs; and focus on community upsides and the social aspect to allay one of the older generation's greatest fears – loneliness. None of us wants to admit that we're getting older. It's a psychological PR minefield, but one angle that local authority developers can and should push is continued quality of life in our later years.

Box 6.3 Jeremy Porteus, chief executive, Housing LIN: comment

Local authorities are proactively working to release land with the capacity to develop at least 160,000 homes. And just to meet current and future housing need for mainstream housing, their representative body – the Local Government Association (LGA) – has called for a significant increase the supply of new homes. Most experts believe that we need to build upwards of 250,000 new homes per year.

A report by the Housing Learning and Improvement Network for the LGA revealed that the demand for housing traverses all age groups and that there is a similar pressing need for specialist housing for our ageing population, with 400,000 purpose-built homes to Housing our Ageing Population Panel for Innovation (HAPPI) design principles required by 2030. The LGA has called for a 'residential revolution' in which we meet the housing and lifestyle choices of both our current generation of older people and future generations.

And, as highlighted in the recent government guidance, local authorities play a key strategic role in planning housing for older disabled people. For example, working with social landlords, ALMOs and for-profit operators and developers to meet an increasingly diverse range of tenure options and accommodation types. This includes working with specialist and general needs housing associations to develop HAPPI influenced housing designed with ageing in mind, whether sheltered housing or extra care housing or an over-55s retirement market. However, there is also a growing trend for local authorities as housebuilders to set up their own local housing companies to boost supply for much-needed new homes, including filling a gap to meet the unmet needs for older people's housing.

Box 6.4 Gwyn Owen, head of Essex Housing, Essex County Council: comment

Essex is one of the largest counties in England. The county has a diverse range of local authorities comprising unitary, district, borough and city councils as well as town and parish councils and the County Council. Despite this, the public sector in Essex has worked closely together to try to address housing needs across the county and Essex Housing has been one such initiative to emerge from this collaboration. The Essex Partnership agreed to form a new vehicle that would work on behalf of the public sector, taking an innovative, bespoke approach to realise the financial and social value of development. Essex Housing was established on the basis that our approach to development needs to be inclusive of everyone – residents, private and public sectors who have a stake in any new development.

We seek to balance general needs housing with affordable homes and specialist accommodation for people with additional needs. At any one time around 40% of the units we are bringing forward across our portfolio are either affordable or specialist. While there can be a temptation for local authorities to simply focus on financial returns when considering development, we have managed to balance the delivery of general needs, affordable and specialist units to maximise benefits to Essex as a whole.

Our new and refurbished properties are not only developed to a good, aspirational specification, often with underfloor heating, solar panels, good quality green space, and electric car charging points, but are also marketed thus. It is important to note that we are very careful to ensure we maximise value for money for the Essex taxpayer – while we ensure we provide good quality units, we are clear that every penny counts. For example, although each of our show apartments has a totally different feel depending on the target audience, where possible we seek to reuse furniture and other fixtures from one new development to the next to keep costs to an absolute minimum.

The real difference that Essex Housing brings is that we've maintained the very best of local government culture but added a commercial mindset. We're passionate about helping to meet current and future housing needs. It is clear that to achieve this we must be ambitious and go further than simply seeking to meet our statutory obligations with regards to housing.

Conclusion

The UK, already in the grip of a major housing deficit, is starting to struggle with a fast-ageing population. This challenge, and others, will only be tackled by a strategic approach to building new housing by both the public and the private sector. This may consist of organisations working in isolation as competitors or in partnership, using delivery models previously unseen. Either way, the communications to support this must evolve. As shown in the case study example, there is much that social housing providers can learn from those who have longer-term experience of providing homes for both private and public sectors. The major PR challenge for local authority developers is in defining their changing offering and USPs, communicating them in a more commercial way, and shedding their previous public sector identity, both internally and externally.

Notes

1 The Decent Homes Programme was brought in by the 1997 Labour administration to provide a minimum standard of housing conditions for those living in public sector (council housing and housing associations) homes through the Decent Homes Standard. It required local authorities to set out a timetable under which they would assess, modify and, where necessary, replace their housing stock according to the conditions laid out in the standard.

2 www.architectsjournal.co.uk/news/news-analysis-why-are-councils-setting-up-private-companies-to-build-homes/10017168.article [Accessed 15 September 2019].

3 www.independent.co.uk/news/uk/home-news/housing-homeless-crisis-homes-a8356646.html#targetText=Groundbreaking%20research%20by%20Heriot%2DWatt,target%20of%20300%2C000%20homes%20annually [Accessed 15 September 2019].

4 www.theargus.co.uk/news/17736856.worthing-council-hopes-to-build-150-ikea-eco-homes/ [Accessed 14 October 2019].

5 www.alzheimers.org.uk/about-us/policy-and-influencing/dementia-uk-report#targetText=The%20number%20of%20people%20with,over%202%20million%20by%202051.&targetText=This%20equals%20one%20in%20every,aged%2065%20years%20and%20over [Accessed 15 September 2019].

6 www.ageuk.org.uk/globalassets/age-uk/documents/reports-and-publications/later_life_uk_factsheet.pdf#targetText=By%202030%2C%20one%20in%20five,%2B%20(ONS%2C%202017b).&targetText=The%2085%2B%20age%20group%20is,)%20(ONS%2C%202018k) [Accessed 15 September 2019].

7 www.nhbcfoundation.org/publication/beyond-location-location-location-priorities-of-new-home-buyers [Accessed 15 September 2019].

7   Promoting student housing

Maintaining value and predicting the future

Henry Columbine

Introduction

The purpose-built student accommodation sector in the UK has been transformed beyond recognition over the last 20 years.

The idea of undertaking sophisticated marketing and PR campaigns for the student digs that most Baby Boomers or Generation X[1] experienced would have been laughable. But developers and investors have seen an opportunity to profit from revitalising an outdated offering and updating it to reflect modern students' demands.

Similarly to the build to rent and hostel sectors, which have followed suit, student accommodation has been transformed from a murky world that relied on young people having limited choice on a limited budget to one that is centred on the customer. In doing so, the sector has gone from being an alternative asset class to one that has started to attract institutional interest and is now almost mainstream.

Interest has been driven by a combination of factors, including government policy to encourage more young people to go to university, growing economies throughout the world – particularly in Asia – and rising student mobility, as well as a change in attitude among students, who are increasingly focused on experience and are therefore willing to pay more for a better standard of living.

Communication for student accommodation brands can involve a number of different but complementary strategies. Most brands will have a requirement for corporate-level B2B communications, local stakeholder engagement (for example with universities, local councils and local communities), and consumer-focused brand-building and asset-level PR to support reservations and rental take-up.

Communications considerations

Communicating around student accommodation has always been likely to involve certain challenges:

- Student accommodation developments are not always welcomed by local communities, who might be opposed to student housing being created close to their homes because of preconceptions about residents' behaviour.

- There is a need to reach diverse audiences: the brand may need to communicate with investors, agents, partners, students and their parents, and the local community around its developments.
- The end users' priorities are constantly changing: much is currently being made of the property industry's adaptation to millennial consumers who are more design conscious, connected and experience focused than ever before. But student accommodation providers need to be one step ahead: this sector is generally a young adult's first interaction with the property industry. To communicate effectively in this sector, we need to monitor and predict, rather than react to, changing values and demands.

As the student accommodation sector in the UK matures, a number of further PR challenges are coming to the fore:

- The majority of new student accommodation developments focus on the luxury end of the market. Similarly to how the London residential sector was criticised for catering excessively to high-end overseas buyers, student accommodation providers should be ready to respond to accusations of not providing sufficient affordable stock. With UK student loans expected to grow to over £1 trillion over the next 25 years[2] and people struggling to get on the housing ladder more than ever before, student debt levels could cause a backlash against schemes that are perceived as being overly expensive and harming young people's financial positions from an early age.
- A high quantity of stock has been added to the market over the last ten years and is starting to age (some of it not very well): whereas most student accommodation campaigns to date have been about launching new brands and new schemes, requirements are now likely to shift as brands mature and schemes require more work to drive occupancy levels.
- Political factors such as the UK's relationships with other countries and changing government policies affecting higher education, coupled with the rising cost of being a student, continually affect student numbers. The possibility of a drop in demand for student accommodation, however, does not seem likely: figures from UCAS show a rise in university applications between 2018 and 2019, with a record number of applicants from outside the UK, mainly driven by China.[3] In March 2019, the UK government announced intentions to increase international student numbers in the UK by more than 30 per cent (600,000 students) by 2030. It remains to be seen whether these ambitions will be realised, but as a location with 11 of the world's top 100 universities,[4] the UK is likely to remain an attractive place to study.

- With more investors and developers having entered the market, there are now fewer opportunities, as well as more competition for well-located development sites. Campaigns targeting agents, landowners and universities to ensure that student accommodation developers have visibility of the best sites – ideally off-market – are therefore becoming more important, and investors are increasingly targeting less mature markets overseas.

Box 7.1 Amro Real Estate Partners: European expansion

Having established itself as a specialist student housing and build to rent investment and asset management business in the UK, Amro Real Estate Partners recognised that there was a significant opportunity for purpose-built student accommodation in southern Europe. In May 2018, at an early stage of its European roll-out, a communications campaign, outlining the company's intention to expand across Iberia and announcing its appointment of a leading global advisory company to seek an institutional joint venture partner, targeted the pan-European property media, helping to bring forward interested parties and those with relevant sites to sell. By June 2019, the company had acquired three assets in Spain: in Málaga, Granada and Seville. Announcements around deals helped to demonstrate Amro's successful execution of its corporate plan and to remind key target audiences of its continued expansion in the region and its ongoing interest in additional sites.

Developing communications strategies

Establishing objectives and scope

Any PR and communications campaign should have as its goal the fulfilment of one or more fundamental business objectives. Raising the profile of a business or development is not enough: it is important to understand from the outset whether the campaign needs to bring forward development opportunities or potential partners, to help secure planning consent, to attract residents to a specific scheme, or to build brand equity to encourage future demand and approaches and enable higher pricing. Often, a student accommodation brand may require a campaign that mixes several objectives, or that transitions from one specific aim to another.

Understanding key moments

Working around set milestones is often the most sensible way to create a communication plan. Consider elements such as site acquisitions, planning consultations, submissions and committee dates, construction events (such as start on site, topping out and completion), launch to the market and first residents moving in. Overlaying corporate news such as senior appointments, sales and funding, or investment activity will ensure that communications activities are coordinated and mutually supportive.

Establishing audiences and stakeholders

We have already touched upon the likely diversity of audiences for campaigns in this sector. Even on a relatively simple campaign to support lettings, it is important to remember that audiences may well be international and that messaging also needs to appeal to parents, who will often be the ones footing the bill, as well as end users.

Identifying other stakeholders is also key. Strong relationships with the communications teams at local universities will be helpful, as coordination over opportunities, events or media enquiries may often be required. Similarly, contractors, management companies, investors and local authorities may need to be involved.

Research

Regardless of whether the PR campaign is corporate or consumer focused, understanding audiences and their priorities is essential in order to devise messaging that resonates with them.

Too often, this stage is culled from the process due to a perception that research is too expensive or because of a perceived lack of time. While there is an element of truth in saying that no research is probably better than bad research, there are multiple ways to generate insights that help inform a more effective campaign, not all of which involve nationally representative surveys. For example, using a small sample size to have a focused (qualitative) discussion on what investors in student accommodation want, what their concerns are and why they are still investing in the sector may not result in statistics you can use in the national press, but would give a very good indication of current investor sentiment and provide a steer on the key elements that the messaging of any campaign would need to touch on.

Equally, understanding the brand you are representing is vital. Too many corporate campaigns are built on assumptions and without any understanding of the baseline against which success should be measured. A perception audit, which involves speaking to a representative range of stakeholders, including end users, investors, other stakeholders and the media, can prove insightful in understanding key strengths to be emphasised, weaknesses to be countered and misconceptions to be corrected.

From a consumer perspective, research is arguably even more important. As new waves of students enter the marketplace, understanding how the latest generation lives and wants to live, their attitudes and values, and how a campaign can tap into these elements could make the difference between messaging that speaks to young people effectively and one that fails to hit the mark or appears outdated. Students and school leavers are a time-rich, cash-poor demographic that is generally bursting with fresh ideas, so engaging them in research tends to be both fruitful and inexpensive.

On scheme-specific campaigns, understanding the local market and competing brands and buildings will also be vital in order to create messaging that differentiates the product you are promoting from its competition. Engaging with agents or sales and marketing directors, as well as local and internet-based research, can be helpful in this regard.

Box 7.2 Uliving: research and implementation of research findings

Recognising that being in touch with the current sentiment of students was critical to success in the student accommodation sector, Uliving and Bouygues UK engaged with University of the Arts London Central Saint Martins on an initiative called Beyond Dwelling. Involving the students of its BA Architecture: Spaces and Objects course, the project saw groups of students invited to consider the future of student accommodation in London and to submit proposals for new schemes. The project culminated in an exhibition, displaying a series of fresh ideas and highlighting that students' expectations went far beyond provision of accommodation, with social, cultural and community aspects also being incorporated into their designs, and every entry including sustainability in some form.

The winning proposal was designed to bridge the gap between student and professional life, enabling the growth of an open-source community among students of the arts and young artists. Composed of two buildings, the proposed scheme would provide a student living space linked via a courtyard, three main bridges, and a promenade of large balconies providing collaboration and exhibition space to a building offering plug-in affordable studio space. Despite being put together by a group of technologically connected students in the mid-2010s, the ethos behind the design was to restore physical and visual interaction in a world where students spend increasing time indoors, interacting virtually.

The project helped to inform Uliving's future developments as well as its communications activity, while also positioning the company as a forward-thinking brand that puts students at the centre of its work.

Language and messaging

In a sector that often involves multiple and very different stakeholders, messaging will generally need to be carved out according to what is important and relevant to each audience. It is also worth investing time upfront to create a coherent brand language that, while being subtly different in tone for different audiences, is not so different that it appears inauthentic or inconsistent.

Communicating with younger audiences can be something of a minefield: one wrong word can undermine a whole campaign. Senior property executives are rarely in tune with what will appeal to a student, so managing clients (whether internal or external) carefully, pushing back where necessary, referring to research, and ensuring that messaging and language are tested on focus groups before being rolled out can help avoid pitfalls and ensure that messaging is chosen based on how it is received by its intended audience rather than on others' opinions. A few general guidelines follow:

- Don't patronise: these are young adults, not children.
- Don't tell them what to think: using subjective terms such as 'great' or, worst of all, 'cool' looks desperate; objective information, implication and independent endorsement will be far more effective.
- Don't use jargon or words that will be alien to your audience; while a property developer may talk about 'units', you should be talking to customers about 'rooms', 'apartments' or 'homes'.

Specific messages will be defined by the product or brand itself, but think about what the brand or product is offering that is different, unique or new. If aiming to promote a new scheme, think about what it offers in terms of product, lifestyle, amenities, technology and community. Does it bring anything new to an area or to the sector? Is it the biggest, best, first, most advanced, best designed? All messaging should be backed up with proof points.

Communications materials

This is a young, dynamic and visual sector, and B2C – and, to some extent, B2B – communications materials should reflect that. Consider how best to get your messages across, as often this will be done more effectively through video and photography than through wordy materials. Use of third-party endorsement – whether from students, universities or other stakeholders – will often also be more convincing than materials with an overly corporate, first-person stance.

Establishing relevant channels

While reaching investor and trade audiences can be done relatively effectively through news flow and thought leadership content within the media and through industry events, students and younger audiences can be harder to reach. Research from Ofcom has found that, unsurprisingly, the internet is the most popular platform for news for 82 per cent of 16- to 24-year-olds,[5] but with a new student accommodation block unlikely to hit the national news headlines, it is important to think about how this elusive audience can be targeted effectively.

Social media is an obvious answer and should form part of any campaign focused on reaching a younger audience. Statistics from Pew Research Centre highlight that Instagram and Snapchat are the most popular social media or messaging platforms among younger age groups. Whereas 51 per cent of 13–17-year-olds use Facebook and 32 per cent use Twitter, 72 per cent use Instagram and 69 per cent use Snapchat. Facebook might, however, also be worth considering for its broad popularity and for reaching audiences such as parents, while, from a corporate and investment perspective, LinkedIn may be an appropriate channel.

With Gen Z being the first generation to have grown up in the post-dial-up-internet age, social media activity focused on end users needs to be carefully considered. For example, there is evidence to suggest a lack of trust in influencers,[6] so simply paying for positive promotion is unlikely to cut it with a digitally au-fait generation. Too often, social media is seen as a sales tool, whereas it should be regarded as a brand-building tool: activity needs to be authentic, creative, sustained and subtle to be successful. People will choose to engage with interesting, thought-provoking content; they are unlikely to choose to listen to sales messages.

Box 7.3 Nido student accommodation: use of YouTube

Nido's YouTube channel acts as a repository for a collection of thought-provoking short films that create a strong perception of Nido as an international, creative and aesthetically focused brand. Its videos include residents talking about their lives and successes – including a photographer and an international ballerina – and lifestyle-focused tours of areas where the company's schemes are located. The properties themselves are barely mentioned, if at all; the focus instead is on creating content that matches the aspirations and values of residents and potential residents.

Figure 7.1 Most-used words associated with 'student accommodation' on Twitter (1 January to 1 July 2019).

Figure 7.2 Most-used emojis associated with 'student accommodation' on Twitter (1 January to 1 July 2019).

Figure 7.3 Most-used hashtags associated with 'student accommodation' on Twitter (1 January to 1 July 2019).

Events, partnerships and collaborations

Fortunately, the student community is a hotbed of creativity and one that is generally open to attending free social events. Think about how events and social media strategies can intertwine to engage with both local and geographically distant audiences, using digital channels to drive attendees to an event and using events to provide content for digital channels. Use research to understand the type of events that will appeal and will fit with the brand and product you are promoting. Anything from a talk with an interior designer on how to decorate a room with temporary furnishings on a shoestring budget, to pop-up bars and street-food markets, through to outdoor film or sports screenings, fashion shows or exercise classes could attract footfall onto a scheme and create a buzz among the student community.

Box 7.4 Unite: the Common Room

The Common Room is a section of Unite's website that provides advice pieces, features and lifestyle content tailored to its student residents and those about to start university. This microsite not only helps with search engine optimisation; it also positions Unite as an expert brand on student living and creates a feeling of community among residents – some of whom share their experiences through first-person articles.

Article topics include a rundown of useful apps for student life, must-have kitchen utensils, dealing with loneliness and mental health, LGBTQ+ perspectives and how to budget effectively.

Figure 7.4 Unite: The Common Room.

Also, think about brand collaborations. If your scheme includes a gym, could you tie up with a local DJ to provide a workout soundtrack one evening? Could you collaborate with local bars, cafes or independent shops to offer discounts or perks for your residents, demonstrating your development's integration with the local area and helping to align your brand with those that are liked by your target market?

In schemes that are already occupied, bringing together existing students can have a powerful effect not only in keeping existing tenants but also in attracting new ones through demonstrating a sense of community.

Providing protection

A PR and communications campaign in the student accommodation sector should focus not purely on promotion but also on protection.

This is a subsector of the property market in which having a comprehensive crisis communications plan is particularly important. This should outline the most likely issues that could arise – from bed-bug infestations to accidents, through to kitchen fires and noise complaints – and provide a protocol for dealing with them, contact details of relevant parties, and pre-prepared holding statements so that any media enquiries can be handled swiftly and any brand damage can be minimised.

For many residents of student accommodation schemes, this will be the first time they have lived away from home. Not only does this make accidents and incidents more likely; it also makes dealing with any issues quickly and satisfactorily all the more important. With many parents paying rents and concerned for the welfare of their children, communication around any issues needs to be prompt and reassuring. Only when people feel they are not being listened to do they tend to make their concerns public, escalating them to the media, social channels or review sites. Health and safety of what is a vulnerable demographic needs to be a top priority.

Attention needs to be paid to the student timetable, too. If maintenance works need to be undertaken during a period when students may be revising for exams, for example, consider how this can be communicated to them and show that you aim to keep disruption to a minimum.

Many issues will be at a local level, so fostering a positive relationship with the local and regional media, sponsoring local events or working with a charity in the area can be helpful from the start to ensure that the local community feels that the development has a positive impact on its environment.

Measurement

As with any campaign, the focus in terms of measurement should be on impacts rather than outputs. KPIs should be agreed at the outset and might include changes in brand perception and net promoter scores; positive mentions or engagement on social media; increases in web traffic, social media views or followers; letting rates and leads; or number of approaches from potential corporate partners.

Again, some element of research is likely to be useful here, and qualitative research should not always be shunned in favour of something that can be turned into a number. For example, a perception audit carried out before and after a campaign may not provide figures but will give an idea of the extent to which the dial has been moved.

The future

The student accommodation sector has seen exceptional levels of change over the last 20 years and remains one of the most dynamic areas within the real estate space.

In the UK, the market will only get more competitive: there are now more players than ever; customer expectations are at an all-time high, it is becoming more difficult to offer something new and different, and increased stock and ageing buildings are likely to bring new challenges to communicators within the sector. We are likely to see more investors and developers turning to Europe, where the purpose-built student accommodation sector is at a much earlier stage, in order to maintain returns. Being able to provide strategic communications at an international level is therefore likely to become increasingly important.

This is a sector where there are lots of opportunities to be creative within the communications plan, with budget normally being the only sticking point.

As disciplines increasingly converge, with the lines between PR, marketing, sales, HR and CSR blurring when it comes to certain areas, it is important to take a coordinated approach. Sponsorship of university teams, festivals or sports events, CSR work in the student or local communities, virtual reality tours, and guerrilla marketing such as poster campaigns and clean or chalk graffiti may not necessarily fall within the remit of a communications team but could be useful additional dimensions or proof points to support a campaign.

In an area where the target market's attitudes and values are perhaps more difficult to ascertain than any other within the property sector, the best campaigns will remain those that are based on insights into the target audience, engaging with them through relevant and thought-provoking content.

Notes

1 For the purposes of this chapter, 'Baby Boomers' are defined as those born between 1946 and 1964, 'Generation X' comprises those born between 1965 and 1980, 'Millennials' are those born between 1981 and 2000, and 'Generation Z' includes anyone born after 2000.

2 Coughlan, S. *Student loans 'rising to trillion pounds'* (2019) [online]. BBC News. www.bbc.co.uk/news/education-44433569 [Accessed 5 July 2019].

3 UCAS. *January Deadline Analysis Report* (2019) [online]. UCAS. www.ucas.com/file/214311/download?token=IWBoM4PE [Accessed 5 July 2019].

4 www.timeshighereducation.com/world-university-rankings/2019/world-ranking #!/page/0/length/100/locations/GB/sort_by/rank/sort_order/asc/cols/stats [Accessed 5 July 2019].

5 Jigsaw Research. *News Consumption in the UK* (2018) [online]. Ofcom. www.ofcom.org.uk/__data/assets/pdf_file/0024/116529/news-consumption-2018.pdf [Accessed 5 July 2019].

6 We Are Social. *We Are Gen Z – Their Power and Their Paradox* (2019) [online]. London: We Are Social, pp. 51–56. https://wearesocial.com/uk/wp-content/themes/wearesocial/attach/genz/We_Are_Gen_Z-Their_Power_and_their_Paradox.pdf [Accessed 5 July 2019].

8 Promoting high-end property

Where less is more – the art of communication at the top end of the market

Tania Thomas and Henrietta Harwood-Smith

Setting the scene

The prime property market in the UK is undoubtedly impacted by political change. At the time of writing, the 2014 increase in stamp duty for properties priced over £1.5 million has forced both investors and owner occupiers to re-think their house purchasing decisions – often it makes more financial sense to improve than to move or to invest in different assets. Brexit, too, has caused uncertainty, with many investors from across the globe sitting on their hands pending decisions being made.

The statistics show that London house prices fell at the sharpest rate in a decade (4.4 per cent) in the year to May 2019.[1] But lower house prices and the combined macro implications meant that as the pound continued to dive against other currencies, there was a silver lining: property investment in the UK represented excellent value. Knight Frank data shows that investment into prime London properties by American purchasers, for example, doubled in the first part of 2019 compared with 2018.[2]

So, writing in mid-2019, it is clear that we're in unprecedented territory and that developers and agents at the top end of the residential property sector are having to think out of the box when communicating with their target markets.

Communication tactics

Gone are the days when promoting a high-end property development in Mayfair simply involved placing an advertisement in the *Financial Times* and expecting the phone to ring off the hook.

Potential purchasers at the top end require thoughtful and more subtle 'courting' from a range of marketing channels to increase brand awareness, and we increasingly target different international territories. Ultra-high net worth individuals tend to lead transient and time-poor lives – they could be in London for business before travelling to New York for an appointment and then back for a holiday in the South of France all within a few days, so extra effort needs to be made to capture their attention. High net worth individuals also value their privacy and security hugely, so discretion is paramount. They want to feel that they are reading and finding out about opportunities that others may not know about, which presents a challenge for a PR campaign. Often, it is as much about keeping a story out of the press: knowing which aspects to talk about and which to hold back.

A well-considered, sensitive and targeted approach to PR must be taken, and the strategic international reach of a high-end property campaign across a range of influential titles is crucial to success.

Importantly, PR consultants must have established and trusted relationships with the relevant media that the brand needs to be speaking to, and must understand the target audiences and landscape that they orbit. They must also have the creativity to guide brands in setting themselves apart in thoughtful, inspiring ways – following a 'cookie cutter' approach will not only fall flat with the media but will be unlikely to capture the attention of key target audiences.

Agreeing the strategy

The most successful property development campaigns are those where the PR teams have been involved from the acquisition and planning stages, working closely with sales and marketing teams, considering the wider placemaking messaging and overall objectives of the campaign. Early involvement enables the PR team to input expertise to guide and navigate campaigns and to ensure that the proposition (e.g. the features of the property development) provides strong content to support the key messages required to fulfil the business objectives from start to finish.

An integrated approach

While PR and marketing used to exist as two separate entities, to be successful today, these two strands should work succinctly and in tune with one another. This ensures that the campaign is joined up, whether it's a story running in the media or a direct mailer and advertising campaign. All these touch points should work harmoniously to drive consistent messaging.

The fast-changing world of digital media has blurred the lines more and more, and increasingly a story will break online first before it gets into print. In mainstream public relations, traditional forms of PR (such as print media) are taking a back seat, with digital becoming the primary source. This means that today many PR professionals are looking towards blogging, influencer relations, social media management, search engine optimisation (SEO) and contributing content to other platforms to meet the needs of their clients.

However, the luxury sector is the one area where print media (usually accompanied by an online version of the same piece) still reigns supreme and has an incredibly important and relevant place. There is an immediate association with quality and luxury that comes from appearing in a glossy magazine or the quality pink printed pages of the *Financial Times*. Magazines have a longer shelf life and can be read and enjoyed in varying locations. Having time to sit and read a printed article in a magazine today is a true luxury, so such articles can create impact. This means that campaigns targeting ultra-high net worth individuals should always have a strong print media component and focus.

The key channels that tend to have the most traction, when carefully timed and managed as part of an overall communications strategy and budget, are as follows (and note that they all need to be considered harmoniously):

- traditional PR and print-focused media relations
- advertorials
- TV and radio mentions where appropriate
- advertising
- social media: case by case depending on campaign – in particular, Instagram, Twitter and LinkedIn.

Choosing focus territories

The PR campaign needs to be closely aligned with the commercial sales objectives. When publicising a new high-end development, for example, by working closely with the real estate sales teams, the communications experts can identify focus target hubs. This will need to be adapted and evolved as the sales process gets under way – it may be that the majority of interest comes from Europe or the US; hence, the comms team can build on existing press coverage to further encourage sales.

Leading a campaign from the UK makes sense, as London is still a global hub due to its time zone, recognised as a centre for excellence in education and health, offering leading hotels, restaurants and facilities, and as a financial capital, with a number of leading businesses having a presence here. However, the communication strategy needs to carefully consider exactly which territories will result in positive targeted interest.

Keeping the story fresh

Creativity and originality are key to the strategy to ensure that a brand stands out.

The lifecycle of property developments (and traditional sales) at all levels of the market is such that a PR and communications campaign can often run over a few years. After the buzz of the initial launch, it's crucial to consider how to keep the story fresh in the eyes of both the media and key target audiences. It's important to have a robust and clear yet flexible strategy in place, given that timings and delivery can change. It helps to have clear 'PR moments' marked out in the plan, which will create appropriate news hooks, not giving everything away too soon, so as to 'keep powder dry' for impactful story-telling later into the campaign.

Awareness and 'PR moments' can be created and elevated through additional creative brand partnerships that resonate with high net worth audiences.

Computer-generated imagery (CGI) is useful but needs to be of the highest quality to create cut-through. Such images are a worthwhile investment and can really bring a development to life when there is nothing physical to see. It's not enough, however, to keep rolling out the same images. Importantly, now that digital media forms such a central part of any campaign, journalists are able to see exactly what others have featured. Hence, fresh imagery at various key milestones, such as the launch of a new show apartment, can be impactful.

From an ideas perspective, an overall communications strategy needs to be created to ensure that the various milestones form tangible hooks for the media. Post launch, carefully timed hooks have to be sensitively plotted to create cut-through without over-saturating press interest.

Key messages

Developers and agents at the top end need to carefully consider the types of topics and issues that will resonate with their target audience.

Topics and messages that often strike a chord are:

- sustainability
- design excellence
- health and wellness
- craftsmanship
- thoughtful sourcing
- art and collectables
- privacy and security
- diversity
- refinement and modesty
- philanthropy
- innovation and tech.

'The psychology of luxury is changing. More thoughtful areas of luxury are being explored including simple human kindness.'

Kate Reardon, Editor of *The Times* Luxx supplement[3]

Creativity: a focus on design

While architecture is important in all aspects of property PR, it is a particularly pivotal cog at the top end. The intricacies will likely be of huge interest to potential purchasers from both an interior design and a styling standpoint. The choice of architect can make or break a campaign. In years gone by, many developers were happy to merely dress a show apartment without going into too much 'story-telling' detail or challenging their designers to carefully consider each element of their sourcing. Now, it's the attention to detail that tends to generate the column inches. Whether it be the lighting that positively impacts circadian rhythms or the dining room table handmade by experienced craftsmen and women in the UK, it's the thoughtful details that help to create the full picture. Purchasers buy into the design experience as much as the pure bricks and mortar.

Box 8.1 Great Portland Estates, Rathbone Square: promoting design

Great Portland Estates had successfully sold the residential units at Rathbone Square, London and let the commercial space to Facebook but intentionally held back the two penthouses to market once the development was complete. Both were on the market for in excess of £7 million. Maison Communications recommended working with a British-based leading designer known for their attention to detail and versatility (which complemented Great Portland Estates' own values) and brought a few options to the table. The challenge was to both 'placemake' Rathbone Square, Fitzrovia as one of central London's leading prime residential developments (in a highly competitive market) and to ensure that this offering stood apart. For Great Portland Estates this was to be a celebration of the completion of the building, and the design needed to be exceptional and act as a showcase.

Echlin London, a member of Walpole British Luxury brands, was chosen because of its commitment to craftsmanship and quality design.

In addition to the design, key topics that formed the backbone of the campaign were:

- British heritage and Echlin craftsmanship underpinning the thoughtful sourcing;
- brand partnerships via Echlin's network of craftspeople, which included the British-based ceramic artist Lauren Nauman;
- art and collectables with relevance to the local area and history;
- Great Portland Estates as a trusted and leading London developer with a commitment to regenerating pockets of the capital to create a long-lasting legacy for future generations;
- the wider placemaking story and proximity to Crossrail;
- Great Portland Estates leading the regeneration of the Eastern end of Oxford Street;

- the story-telling behind the design and smaller details such as the cosmetics fridge in the master bathroom;
- open plan living;
- London's best penthouse views.

By communicating carefully thought-out messages in a range of international territories, a positive sales outcome was achieved.

Research

Even the most wealthy global property purchasers want to know that they're making a savvy investment choice. They might be buying either for themselves, for their children or purely for investment purposes, but local market insights and wider research are likely to underpin their decisions. Developers in particular often work with their agent's in-house research teams to create micro market insight reports. However, working with external research houses or brand partners can be particularly effective to create a 'one of a kind' insight report that will further outline the benefits of a purchase. The media frequently picks up on fresh statistics that help shape a wider story.

Box 8.2 The Modern House: using research

Maison Communications was appointed by specialist estate agency The Modern House to raise the profile of the company and its successes to date across consumer and industry audiences, positioning it as the leading specialist estate agency for design-led properties.

Maison worked with The Modern House to get the most out of a research report, which it produced in collaboration with Dataloft, identifying the transformative effect of design on the value of homes in London. Maison advised on how to get the most out of the report and how to package the pitches to media to secure the strongest coverage across leading trade, lifestyle, design, architecture and international press. There was a good bank of case studies and evidence to back up the key messages.

Over an eight-week PR campaign, coverage of the research reached over 800,000 readers. In response to the brief, the research report was featured across a number of top tier national UK titles, along with a few bonus international pieces.

Brand collaborations

When considering the overall picture of how to reach target audiences, brand collaborations can be a particularly useful tool. Tapping into the broader audiences of symbiotic brands can be extremely impactful across all strands of the marketing mix.

A potential purchaser of a high-end property may also be following complementary brands in the worlds of fashion, design, cars, planes, yachts, food, health and wellness, and culture. By joining forces on an element of the campaign, such as piece of joint research or the launch of a new show apartment with a well-known high-end fashion designer, this can help to keep the story fresh and to build momentum. It can also be particularly impactful from a social media perspective to build a following.

Box 8.3 Brand collaboration in placemaking

Martin's Properties was founded in 1946 by the Martin family. Since then, the family has created one of the largest property portfolios within the Royal Borough of Kensington and Chelsea, but until recently has remained relatively under the radar. 2018 marked the first time in the company's history of a non-family member being appointed managing director. Under the new stewardship it is diversifying its portfolio across London and the south-east of England to include 80 residential units and 63 commercial units, as well as introducing innovative strategies to revive and reinvigorate the locations where it develops and manages assets.

In line with this, the role of marketing and PR has been crucial to ensure that the brand is recognised and to attract talent, quality tenants and good rents. A large proportion of the estate includes properties on and around the King's Road. Martin's Properties attracted the now world-famous Chelsea Ivy to the King's Road in 2015, which has been a game changer for that area of the road.

Maison Communications' role included overseeing the communication of the new Martin's Properties brand, promotion of a new website, introductions to key press, acquisition and sales announcements, retail lease announcements, profile raising features for the senior team, content creation across all social media channels, a monthly blog covering latest development news and 'meet the tenants' interviews, and introductions to brands and designers for collaborations. A major role for the PR has been placemaking and the repositioning of the King's Road in particular as a residential, retail and leisure destination, and promotion of new developments and tenants alongside impactful brand collaborations.

Recently, the King's Road as a whole has faced challenges, including greater competition from other fashionable streets such as Marylebone High Street, and ambitious placemaking masterplans for new districts of London. This has meant that new destinations have emerged for living, shopping, visiting and working. The King's Road could no longer rest on its laurels and rely on the heritage of bygone years – something needed to change to ensure that it remained a top destination. But the street lacked a single cohesive plan due to its multiple landlords and mixed ownership. Cadogan was already making progress with improvements around Sloane Square at the east end. Martin's Properties had also changed areas in its ownership to create destinations such as the Chelsea Ivy and surrounding buildings, but this needed to join up and filter into Chelsea as a whole. So, Martin's Properties partnered with Cadogan and Halj Group on a new strategy to revive the entire street. The King's Road

Partnership was formed with plans to create more experiences for customers and to improve the public realm.

The new strategy was unveiled in partnership with the Mary Quant exhibition at the V&A in Spring 2019. Mary Quant had caused a sensation on the King's Road in the 1960s when she opened her famous bazaar, and reigniting this buzz seemed hugely relevant. Wink Creative delivered the branding. Major features were lined up and appeared in *FT How To Spend It*, *FT House & Home*, *The Times*, *The Sunday Times* and *Country Life*, along with overseas features and industry trades communicating the messaging of the King's Road revival and highlighting the calibre of the new tenants and collaborations on the street. The press communicated that Martin's Properties had succeeded in attracting a new calibre of occupier to the King's Road, with mentions for the new brands of Sticks n Sushi, Kobox, Schoen Clinic, Matcha & Beyond and the redevelopment of beautiful residential rental properties in and around the King's Road and Radnor Walk. The messaging reflected a new identity for the street, which celebrated health and wellness and support for independent retailers and directional brands, and this has resonated with the media.

This is just the start of an ongoing revival of the King's Road story but demonstrates how brand alignments and major creative initiatives can help an unknown brand stand out in a crowded marketplace.

Evaluation

PR continues to come under increasing scrutiny from clients and investors. Setting targets and KPIs is standard practice. The coverage needs to be accountable and prove its worth. The true measure of a piece of coverage is not in its size; it is in the quality of the coverage.

Other measures (and in some cases a point scoring system is issued) include:

- Quality and relevance of title in terms of reaching target audiences.
- Tonality – is it a positive balanced piece?
- Inclusion of key messages – does the article get the messages across, and if so, how many of them?
- Impact of spread – for print media, where in the paper is it, where on the page is the main focus situated, and are well-captioned images included?
- How is the article received? – for online articles, the comments beneath can be monitored, and the performance of the article and number of hits studied.
- Visits to the website – Google Analytics can monitor the impact of an article on web activity and goal conversions.
- Increased enquiries – where a phone number is given, the number of calls received off the back of people reading an article remains a good measure, although the number of enquiries coming via phone is far lower, with the internet tending to be the priority.

In high-end property PR, less is more, and quality trumps quantity. Every strategy and form of measurement will be bespoke and tailored to the clients' needs and requirements, and needs to remain flexible. The focus is on securing a solid few major editorial features in priority, top tier titles, as opposed to a scatter-gun approach to secure as many column inches as possible.

Conclusion

The luxury market is ever evolving and extremely fast moving. At the time of writing, the overall feeling is of a move away from the ostentatious and a return to kindliness and humility, with carefully thought-out craft and design at its core. The luxury of time is an element that underpins many high-end campaigns – with so many people living 24/7 lives, developers and agents are faced with the challenge of conveying an offline simplicity that creates a sense of harmony. Many brands are now actively avoiding social media 'influencer' marketing tactics, viewing them as insincere. A few key full features with carefully thought-out exclusive media hooks can be far more impactful than a blanket press release approach. There most certainly has been a 'return to basics' from a pure PR perspective – key messages presented in a tangible format in a national print publication, a high-end glossy or an in-flight magazine where people have the time to focus and to read either on their iPad app or via hard copy can often be the most successful.

Overall, the key messages, language, global target territories and brand collaborations as part of the wider marketing strategy will make or break a high-end property campaign. PR and communication as a whole are central to sales success and ultimately the client's bottom line.

Notes

1 Office for National Statistics UK House Price Index: May 2019. www.ons.gov.uk/economy/inflationandpriceindices/bulletins/housepriceindex/may2019 [Accessed 8 August 2019].
2 https://abcnews.go.com/International/us-buyers-share-luxury-london-properties-doubles-year/story?id=63996620 [Accessed 14 October 2019].
3 Speaking at the Luxury Communications Council (LCC) Forum, November 2018.

9 Promoting interior design

Visual creativity and online media – a PR match made in heaven

Tricia Topping

Introduction

Interior design has long been regarded as the Cinderella of the property world, seen as a cottage industry staffed by people more interested in fabrics than in keeping to a budget. With a growing interest in design from enlightened consumers, things are changing, and good interior design businesses are rising in prominence. These are professionally run with a deep understanding of budgets, deadlines and consumer research. In the past, interior design has been a difficult subject to promote, but the big shifting factor is the rise of social media, and Instagram in particular, which as a visual platform is ideally suited to the promotion of interiors. This chapter explains the background to interior design, the difference between the various disciplines, and how social media channels can be used to add real value to a PR and communications campaign.

Box 9.1 The evolution of the interior design business

It is often assumed that Britain's love of visiting show homes at the weekend, known in the industry as 'the carpet treaders', started in the 1990s when interior-designed show homes first became fashionable. But the nation's love of new homes and interiors can be traced back to 1908 and the first Ideal Home Exhibition devised by the *Daily Mail*. Perfect streets of various homes from red-roofed cottages to bungalows were built, attracting thousands of people. In the 1930s the suburbs began to spread, mortgages were becoming available, and more people visited the show to get interior design ideas and to look at innovative new home appliances such as the microwave, which was first launched in 1947 at the show.

Things got difficult in 1970s when a recession and the three-day week meant that many people were more interested in financial survival than in buying or improving homes. There are many examples of houses sold without central heating or kitchens so that they were affordable to cash-strapped purchasers. Curtains and cushions were considered a luxury.

During the 1980s, a decade which was all about money, symbolised by Wall Street and the famous quote 'Greed is Good', television programmes such as *Dallas* and *Dynasty* influenced elaborate interiors. Everything changed

---

with the arrival of Kelly Hoppen and Giorgio Armani, who created the mini-malist grey/beige/cream look that has been embraced in show homes for nearly 20 years. The Ideal Home Exhibition can be credited with creating the aspiration to own the perfect home and became a 'must visit' for home enthu-siasts. In later years, housebuilders committed large advertising and market-ing budgets to creating show homes at these exhibitions with the aim of attracting the 'carpet treaders' to their on-site show homes.

Thirty years ago, furnished show homes were only used for large develop-ments or those that were difficult to sell. In most cases, the chairman's wife or sales director would choose the soft furnishings, while the purchasing depart-ment would select the kitchens and bathrooms and choose where to put the lights.

The gentrification of run-down areas in London led to the rise of show apartments. These areas gave housebuilders an opportunity to buy land cheaply, but the marketing needed to move up a notch, and so did the inte-rior design. In one development of former council offices in south London, eight show apartments were created. The demand to buy these well-priced apartments led to queues around the block, and those waiting were served coffee by girls on roller skates. The rise of sales trips to the Middle East and then the Far East enticed buyers who wanted to see furnished apartments of a style not previously witnessed in London. This demand opened the door to the entrepreneurial interior designers who recognised the new market and the opportunities of working with developers. This was a market that had been previously disregarded by the established interior designers.

Interior design companies now have budgets and deadlines as part of their vocabulary as much as fabrics and tiles. Working as an interior designer for developers is now big business. Clients in the market of multi-million-pound homes demand a recognised *House & Garden* Top 100 designer, and these designers are commissioned for special show homes across London. In complex mixed-use schemes combining residential, retail and office space, architects themselves had been ignored until a well-known journalist, Michael Hanson, mounted a campaign to ensure their recognition, and masterplanners began to demand 'star architects' such as Sir Terry Farrell or John Thompson to help entice buyers to purchase off-plan. In the same way as developers recognised 'starchitects', star interior designers are also now taking a seat at the table. Interior design is now recognised as a key selling tool for any new development, with mainstream interior designers working alongside develop-ers to create beautiful schemes that are far removed from the design horrors of the past.

With space in the capital coming at a premium, the Marie Kondo influ-ence[1] will be felt by developers, through their advisors or their own interests, leading to the commissioning of special pieces from carpets to ceramics to give new homes individuality and creativity. A new breed of developer and housebuilder recognises the value of a well-designed home, and interior designers are starting to feature in the brochures and in award entries. After many years, instead of being an afterthought, interior design has risen to prominence and, in addition to occupying an important place in the lifecycle of property development, is also key to successful property marketing.

Box 9.2 The difference between an interior designer, an interior architect and an interior decorator

To most people, an interior designer is someone who can interpret a client's vision for a home and make the curtains and cushions. For some, it is a hassle-free way of decorating a home. There is a world of difference between an interior architect and an interior designer. The first and most important is that you cannot call yourself an 'architect' without having the necessary architecture qualifications and being registered with the Architects Registration Board, as it is a legally protected title.

An interior architect will consider everything from lighting, heating and ventilation to intelligent use of space. Working in every discipline from commercial to retail and residential, interior architects are worth their weight in gold for their ability to optimise space. It is interesting to note that a Top 10 housebuilder has recently advertised for an interior architect to lead its project, showing that this is central to its development. A prime example of an interior architect was the late David Collins, whose impact on design and colour in hotels and work for superstars, including Madonna, so influenced other designers. His style for hotel interiors is still influencing show home designers today.

An interior designer is concerned with décor and soft furnishing; typically, an expert with an eye for colour, texture and fashion trends. The professionally qualified interior designer with a degree is now a world away from the early interior designers, and perceptions of the industry are starting to change.

The third term, interior decorator, stretches back into the era of Art Deco, when decorators had a foot firmly in the art world. There have been many superb interior decorators in the past, such as the late David Hicks, who inspired the interiors of the swinging sixties. Their concerns are colour schemes, furniture and artwork. Within this section are some of the most influential designers, who recognise trends using their art degrees and experience in colour and design to complement the building's design. They tend to work with high net worth clients with big budgets and design schemes that combine heritage with contemporary style.

The interior design market and PR requirements

Michelle Harris of developers Harris & Home, who has worked on projects in the Caribbean, Switzerland and France and all over the UK, says of the PR function:

> It is very important to work with the architect who will have vision and target audience in mind from the very beginning when they are just sketching out ideas. It is important to use their expertise when thinking of finishes and dressing show apartments as the interiors must continue the journey and style of the exterior and communal areas.

A clever media communicator working with an interior designer must also work with the architect and the marketing department to understand the vision. They will be in touch with trends, target audiences and other demographics and will use their expertise to create a story that works for both online and offline media.

A key to a winning show home is to create the magic that enables the purchaser to imagine themselves living there, and PR's role is to support this. The ultimate accolade for an interior designer is when a buyer purchases all the fixtures and fittings, and PR can help bring this about.

A specialist sector within the interior design arena is rental furniture, and one of the leading rental furniture companies is David Phillips. Managing director Andrew Clark explains that clients are either developers who require rental furniture for show homes or companies providing properties for staff who are relocating. Target audiences are developers and sales and lettings agents: this is very much B2B promotion.

'Our business model is tailored for the property market,' says Clark. 'Our service helps agents to do their job by making it easier to sell or rent properties.'

He says that it's their ability to cover the market from £400 a square foot right the way through to £20 million homes that marks them out, making them the largest supplier of specialist furnishing services in the industry. Retailers have tried to enter the market, but according to Clark, they are more product based and too niche, preferring to focus at one end of the market. 'We're selling a service and not just products,' says Clark. 'We have account managers, interior designers and specialist teams to deal with every aspect of furnishing.'

Clark is also finding that the most innovative designs are in the emerging build to rent space, which has the economies of scale and budget to create anything that the client is looking for:

> We've had bespoke products designed under licence so suppliers can't produce it for anyone else. We designed everything from the apartments to the communal spaces including the residents' lounge, library, cinema, gym and outdoor space. We even had pizza ovens and BBQs made for roof terraces.

It is important when creating publicity for interior design to think strategically and to harness research from the developer, selling agent, interior designer and architect. Media interest is shifting in favour of the interior designer, who has new and exciting routes to market. The top interior designers are now social media influencers, communicators are now publishers, and the developers are also building up followers.

Traditional media versus social media

It has always been difficult to interest the mainstream magazines in interior design for new builds for several reasons, which vary according to the media targeted. Many publications will not consider using interior design schemes from an interior designer specialising in working for developers, as this is viewed as being too promotional and therefore should be paid-for advertising.

There are also issues with photography. The brief to a photographer of a new show home is to sell the property, not to show the creativity of the interior designer. Therefore, a different set of photos, with different object-ives, is required.

Other issues include the reluctance of the interior designer to share their suppliers and the cost of the items. A good interior designer won't want to reveal their trade secrets. Yet journalists writing about interior design want to educate their readers, so helpful sources and tips are essential.

For example, the interior designer may have sourced the fabrics from Chelsea Harbour Design Centre – home to the largest collection of interior design showrooms in the UK and hosting suppliers from kitchens to chandeliers. The Chelsea Harbour Design Centre is a trade showroom, which will not discuss trade prices and instead directs enquiries to a retail supplier. But for publications such as the *Financial Times How to Spend It* magazine, most of their interior features list retails and prices in detail.

High net worth clients who use the finest fabrics and bespoke designs – and therefore have homes that are ultimately 'PR-able' – often insist on a non-disclosure agreement (NDA). This shuts the door on any promotion except in the case of interior designers to the entertainment business who are not bound by NDAs and therefore can offer that added layer of interest for the publication's readership. Peter Sawa from Westgreen Construction, who has had considerable experience in working on multi-million-pound schemes with star interior designers, comments:

> Discretion is paramount with ultra-high net clients. We have created amazing houses and apartments full of bespoke pieces and pushing the boundaries in creativity. NDAs stop us from mentioning the scheme and media coverage would mean we would all [developer, interior designer and PR] lose a client.

Understanding and communicating these issues to the client is an important role of any PR consultant. However, like the fashion world, where bloggers now take precedence over the mainstream media, the media world has changed, with the growth of digital and social media platforms opening up new opportunities for the savvy communications expert.

In its quarterly expenditure report, the Advertising Association and marketing agency WARC calculated that advertising revenues for magazines were expected to fall by 6.2 per cent in 2019, with spend on television advertising rising by just 0.4 per cent.[2] Even blue-chip magazine companies such as Condé Nast have reported the exodus of print advertising, which funds the editorial we all want for our products.

Digital routes to publicity

Advising on promoting websites, a digital consultant and author of several books on the subject of social media, Michelle Carvill of Carvill Creative, advises clients to think of their website as a book on a shelf: to get it to a reader, it needs promotion and publicity. 'Your website is only as good as the publicity it receives,' says Carvill. With so much choice on the internet, it is important not to cloud the issues but to select channels that are appropriate for the business you are promoting. It is impossible to list all the digital channels available for promotion, and there is always the equation of time versus value. Why should you go to the bother of using digital channels? There are several extremely important reasons.

The first, and probably the most important, is that digital is immediate. An image of a development can be posted online and visible to a worldwide audience in minutes. The second most important factor is that the internet is measurable, a subject that has often confused the purchasers of PR services. Analytics and lead tracking software now provides data on who is looking, when they are looking and what they are looking at.

Ideas for new platforms and ways to communicate are created at lightning speed, but for the purpose of promoting interior design it is better to keep to the tried and tested platforms such as Instagram, Pinterest, Twitter, YouTube and Facebook.

Facebook

Facebook has changed the way in which people communicate. Creating a business Facebook page is simple and, when supported by Facebook advertising, can increase brand awareness for an interior designer and their work. Be prepared to invest time in updating pages for you and your clients. A plethora of agencies and freelancers who specialise in this type of work are available to help.

There are very simple rules to follow when using Facebook for interior design, many of which apply to all social media sites:

- Don't hard sell your business, and remember it is not a personal profile.
- Don't post press releases.
- Have great images and videos.
- Promote your Facebook page.
- Read the tutorials that are available.
- Set up Facebook pages for each project.

- Post regularly – but quality is better than quantity.
- Add Facebook social media buttons to your blog and website.

YouTube

Creating videos of interiors should be part of the marketing plan, and this is the place to showcase properties in their full glory. It is a crowded platform, but home sites are a big feature of YouTube. To shine, you must be inventive and creative with your video, and be aware that slick marketing videos tend to be passed over as advertising. The videos need authenticity to get the consumer vote. The biggest rule is 'less is more.' A short one-minute video is ample; otherwise – unless the subject is riveting – your audience will switch off and move on, but remember that your view of what is riveting might differ from their view.

Instagram

This photo- and video-sharing social networking service, now owned by Facebook, is a dream service for promoting the interior designer, as it is all about visuals: it could have been invented purely for the promotion of wonderful interiors. Instagram can create trends though the use of hashtags.

The business service allows small companies to compete with bigger companies, and a simple click on the 'insights' button will tell you the number of impressions, the reach of the image, where new followers are coming from and the number of interactions. It is an excellent way of tracking new followers, finding others to follow or ensuring that you are posting the right image to get attention. It is also possible to promote posts beyond your following by paying a small fee. You can select the age group and range of interests that you wish to target.

Pinterest

This social media platform enables an interior designer to create 'pinning' and 'viewing' boards, focusing on lifestyle and sharing tastes and interest. As a visually focused catalogue of ideas, it is perfect for an interior designer to use to promote their work.

Twitter

This site makes the best use of technology and is the preferred channel of communication for businesses, journalists, celebrities and official announcements. However, it is an over-crowded market, and the risk is that within seconds of posting the information another 1,000 posts will have landed in your target audience's Twitter feed. Promoted tweets can be a good way to target specific groups, including architects and developers who may be interested in sourcing new interior design services.

Influencers

According to data from the Influencer Marketing Hub, the global spend on influencer marketing in 2019 was expected to be $65 billion, up from $1.7 billion in 2016.[3]

This has been described as the Wild West of the internet, where gold is expected in every post. While genuine influencers have very large and loyal followings and can make a big impact on brands, there are also plenty of reports of unethical behaviour. Fake followers can be bought online, and groups known as 'engagement pods' comment on each other's posts to boost engagement. Haphazard pricing and currently no real regulation make this a roller-coaster ride to include in the marketing budget. Influencers can promote brands by looking 'authentic' and avoid ad-blocking. There are several companies that match brands to influencers, and the larger advertising groups are becoming more involved, which should lead to a standardisation in the market. The prices charged by influencers can vary from very low to stratospherically high depending on their level of influence and popularity among advertisers.

As with any market, there are pitfalls in using influencers. When promoting interior design within a show home online, you should:

- find the right influencer and check their following
- select the correct platform
- be clear on your requirements
- adopt a collaborative working approach.

Video-on-demand

VOD allows users to select and watch videos whenever they choose. Vimeo provides a platform to build an audience for interior design and, if the video is good enough, to monetise it. Vimeo can help with video collaboration and distribution. It is a powerful tool, which makes it easy to post videos across channels on Facebook, Twitter, YouTube and LinkedIn. As with most forms of social media, statistics can be analysed in detail.

Traditional routes to publicity

Press exclusives

These are exactly what they say: a story that is exclusive to a specific journalist. If it is for a magazine, you must allow for the lead time, sometimes up to three months, and if, having received an article based on an exclusive from a feature writer, the editor does not like the story, it is binned. Freelance journalists can be good for exclusives, but they often have to sell the story into the publication before visiting a property, and even then it is up to the editor whether the story is published.

Events

According to Alexei Ghavami, marketing manager at Inspired Homes, showcasing a development with an event is great for brand building, but success depends on which journalists attend and making sure that the event fits the location and buyer demographic. Ghavami explains that if the development has shared social spaces such as a residents' lounge or sky terrace, holding events there will show these spaces in use and give buyers a glimpse of what it will be like to live there. 'We hold regular events at our developments as it's all part of the lifestyle that young people using Help to Buy are buying into,' says Ghavami.

Using an event to promote interior design can be particularly challenging. High-end retailers are extremely wary of collaborations, but when a collaboration is achieved, it can be extremely successful, as was demonstrated when London's leading developer St George plc, promoting One Blackfriars development, joined forces with Ralph Lauren and the *Financial Times How to Spend It* magazine to promote a luxury lifestyle through brand association. Working with large retail groups is inspiring but needs considerable time to negotiate to everyone's satisfaction.

Working with the local community

Furnishing and then promoting a development using local artisans, when possible, is a win-win situation for the client and community.

Celebrity endorsement

Finally, ask a rapper to stay in the apartment! The Sugarhill Gang Furious Five needed somewhere to stay for their UK tour, and Inspired Homes let them stay at one of their apartments for free as long as they performed on the development's sky terrace. This created opportunities across all digital platforms using the artists themselves and influencers to endorse the scheme and spread the word.

Box 9.3 Micro apartment at Innova Croydon by Inspired Homes – bronze for Best Interior Design at the *WhatHouse?* Awards 2018: judges' comments

The budget for the interior design of a 30 sq m one-bedroom micro apartment in Croydon was incredible. The developer Inspired Homes gave designer Sarah Akwisombe a budget of just £8,000, which had to include a selection of clothes and VAT, because it had to be attainable for a first-time buyer. The high specification included a great range of fixtures, kitchen appliances from Bosch and Hansgrohe taps, which gave Sarah the basics to create a cool apartment with her micro budget. Clever use of space impressed the judges: the Duffy London coffee table extends to a good-sized dining table and

Pol Potten stools dotted around the apartment provide the seating. A jade green velvet sofa is perfect for watching Netflix with your mates. The bedroom features an Instagram-friendly pink neon light as an alternative to wall art and brings a contemporary edge that suits Croydon to a tee. Sarah also supplied a list and costs for every high street item used to furnish the apartment with a top tips video to assist buyers, a first for an interior design award.

Alexei Ghavami, marketing manager at Inspired Homes, commented: 'Our CEO Martin Skinner is not afraid to push the boundaries in promotion and has given the marketing team the freedom to use alternative strategies that attract a younger audience.'

Box 9.4 Audley Chalfont Dene Retirement Village by Audley Developments – gold for Best Interior Design at the *WhatHouse?* Awards 2016: judges' comments

Instead of a mega mansion, the 2016 gold award went to a retirement development, created in Chalfont St Peter by Audley Retirement for independent retirement living. This was a huge project covering 26,000 sq ft, encompassing 11 apartments and all the facilities to support this incredible village development. The colours and thought that went into creating this interior would not be out of place in a boutique hotel in Mayfair. The design started with a magnificent chandelier in the main hall and the attention to detail with fine joinery and craftsmanship. The bar and lounge areas zing with colour and the restaurant is a delight, with a huge print of a John Singer Sargent painting dominating the private dining area, which can be hired by residents. What impressed the judges most was the idea of an art strategy, where artwork had been sourced on a long-term loan from provincial museums dating from the seventeenth century onwards. These grand portraits and landscapes added a completely different dimension to the development, creating a country house environment where every floor and corner had something interesting to see. The loan benefits the museums, and the paintings are eventually returned, having been restored and loved. The designers, Carol and Kirsty Gearing of Inside Design Company, put enthusiasm, professionalism and passion into the design, and it showed.

Maximising the value of art

Paula Lent of Artmasters is an art curator who was responsible for the combination of contemporary and traditional art in a central London development where the prices reached £50 million. Sculptures and art loaned from art galleries decorated these palatial homes. A sell-out talk by a Lowry specialist was an over-subscribed event as part of a series of promotions that included a series of industry dinners held for property search agents and lawyers, and a Russian brunch linking to the Valentino exhibition at Somerset House.

Art can work at every level. Inspired Homes commissioned several murals on buildings by well-known and local street artists. Street art is associated with change, which is what motivated Inspired Homes when transforming old buildings into innovative new homes for young professionals. The company used Ben Eine, one of the world's most successful street artists, along with local Croydon-based artists Rich Simmons and Art + Believe to attract interest in the show home.

Traditional marketing tools

These ideas have been around for many years and have stayed the course because they continue to work well.

Blogs

Blogs have been popular since the mid-2000s, coinciding with the launch of web publishing tools such as WordPress. The combination of text, digital images and ability to optimise for search engines are the reason for their success. It is estimated that there are over 75 million WordPress blogs now in existence. What better way to demonstrate a step-by step guide to furnishing a show home?

Newsletters

A newsletter can best be defined as a printed report containing news of the activities of a business or organisation that is sent on a regular basis to customers, members or interested parties. While print newsletters are still used, the preference is to capture email addresses via the web (adhering to data protection rules[4]) and to send a regular e-newsletter. The rules remain the same. Primarily, it should provide value and interest to the readers. Newsletters are perfect for promoting interior design in new build properties, offering scope to feature products from kitchens to carpets and how the interior reflects the development.

Mailshots

These are usually very visual and must be no longer than a page. Online, branded emails, which can incorporate videos, are very easy to create, and email marketing platforms such as Mailchimp and Campaign Monitor provide useful statistics to provide information about audiences reached and success matrices.

Awards

Entering awards, whether on behalf of a client or to recognise the work of an interior designer, can add value and prestige to a project. Some interior designers cannot enter awards, as clients' NDAs preclude designers from profiling their properties. In an age when privacy ranks so highly on the list of priorities, many others, including pop stars and wealthy individuals, may also wish to remain anonymous.

There are a few awards available to enter. Each award has specific entry criteria and is aimed at different markets.

Box 9.5 Interior design awards

WhatHouse? *Awards*

These awards have been part of the new-build residential sector for 38 years and attract the biggest names in the industry. They give developers a reason to showcase their products, and there are award categories to suit every type of new build home. The awards have been expanded to include architecture and interior design. The judges physically visit the developments and shortlist the six best entries in each category. The awards are highly competitive and recognised throughout the industry. The award ceremony, which attracts over 1,600 people every year, is a major highlight event in the housebuilding industry and benefits an interior designer and its developer client.

British Home Awards

These 13-year-old awards are run in conjunction with *The Sunday Times* and have several categories that are suitable for interior designers, including Best Interior Design, Individual Home under £500,000 and Individual Home over £500,000. The Best Interior Design category looks for entries that best translate the developer's vision and engage with the buyer.

International Property Awards

These awards cover every sector, including architecture, property marketing, development and interior design. All awards are judged in the UK by a team of over 70 judges recruited from every sector of the market, and award ceremonies are held all over the world. The awards have had a variety of prestigious sponsors, including Rolls Royce and Bentley. The supporting glossy magazine, *International Property & Travel*, can be found in business and first-class lounges around the world. The award entries from numerous international companies are a work of art in themselves, and without a doubt considerable thought and expenditure goes into ensuring that the award attracts the attention of the judges. The interior design entries are often submitted in large purpose-made boxes with sections for the mood boards and fabrics. These international awards offer an opportunity for the developer and interior designer to showcase their work to a worldwide audience.

Interior Design Awards for the trade

The Andrew Martin Awards, International Design and Architecture Awards and SBID awards all have a specific show home and developer section. Winners are generally recognised for their work with private clients. Past winners Kelly Hoppen, Fenton Whelan and Hill House Interiors all have flourishing private client businesses as well as working with housebuilders or in property development.

Box 9.6 How to enter awards

The impact that an award makes when it lands in front of a judge is critical. The judge receives a pile of entries and has to choose which scheme to visit – presentation means everything.

The award needs to entice the judge to open it and read further. Poorly constructed submissions held together with a paper clip that are difficult to read often end up in the bin. Excellent professional photography is a must, along with a full description showing how the interior design meets the client's brief.

If the award is to recognise the interior designer, it must have an endorsement from the client – and if the award is for a show home in a development, it should have a statement from the interior designer concerning the brief.

Before any award entries are undertaken, several briefs need to be written in consultation with the marketing department, including a marketing plan.

- Marketing brief: This explains the award entry and judging criteria to assist the photographer, copywriter and graphic designer creating the submission document. A marketing plan should cover promotion of the award entry, including the social media campaign, blogs, newsletters and press releases. Many companies, large and small, win accolades but fail to use the win to promote the project effectively. The traditional media will not cover other publications' awards, so sending them a press release is a wasted effort.
- Photography brief: This differs from photography of the development and show home. The photography needs to home in on the design with close-ups of the specification, fabrics and finishes. The judges need to get a feel for how the interior designer has interpreted the brief and the products they have expertly sourced.

Copy needs to be crisp, concise and accurate. The judges do not have time to read pages of 'interior design speak' – a picture is worth 1,000 words, and images are the essence of an interior design award.

Understanding the award, researching the judges and ensuring that the submission reaches the destination on time are key. This method of promotion is not cheap but can pay dividends for the interior designer and their client.

Conclusion

Interior design is capturing the zeitgeist of the twenty-first century, in which visual content takes precedence over words. The smartphone reigns supreme, and therefore a focus on digital is imperative. And online or offline, there is no better way to promote a development than to engage with the creativity of interior design.

Keeping ahead of trends is key to successful interior design, and it will be a far-sighted developer who commissions the first sustainable house using pre-loved restored furniture. Homebuyers are consumers at the end of the day, and collaborations with retailers and other consumer brands will help you reach your target audience. With exciting new online brands and powerful social media influencers, this would appear to be the future.

Notes

1 Marie Kondo, also known as Konmari, is a Japanese organising consultant and author. Her best-selling book *The Life-Changing Magic of Tidying Up* has been published in more than 30 countries.
2 Quoted in Walker, J. *Advertising Spend on National and Regional Newsbrands to Fall in 2019 Despite Growing Digital Income, Report Predicts* in *Press Gazette* 4 February 2019. www.pressgazette.co.uk/advertising-spend-on-national-and-regional-news brands-to-fall-in-2019-despite-growing-digital-income-report-predicts [Accessed 9 August 2019].
3 Murphy, Hannah. *Advertising Industry Closes in on a New Target* in the *Financial Times* 12 July 2019. www.ft.com/content/3510eaf0-a3af-11e9-974c-ad1c6ab5efd1# targetText=Advertising%20industry%20closes%20in%20on%20a%20new%20 target%3A%20influencers&targetText=Until%20recently%2C%20influencers %20such%20as,some%20of%20that%20hand%2Dholding [Accessed 2 September 2019].
4 The government's guide to the General Data Protection Regulation is at www. gov.uk/government/publications/guide-to-the-general-data-protection-regulation [Accessed 2 September 2019].

10 Promoting estate agency

Expert insight

Jamie Jago

Introduction

Your home is very likely to be your biggest asset, but selling it is so much more than a financial transaction. You may have lavished love, care and attention on it for decades. Memories are woven into the fabric of the fixtures and fittings you are now sitting down in to discuss with the agents you have invited to pitch for the sale. Who will you trust to market it properly?

The reputation of a company – any company – is arguably its greatest asset, which is why PR is fundamental to success. Clients want to deal with a firm that is well liked and respected, not one that, although well known, is disliked or even reviled. This is especially true in the world of estate agency, which, let's be honest, is not without its critics and detractors.

Whether a small start-up business or the largest, most established international corporation, good PR is more important now than ever to an estate agency. There is so much noise thanks to social media and online, plus the added problem of 'fake news'. But people can still see through the muddled messages and work out what is of value.

Earning third-party endorsements from journalists or bloggers for quality, expertise, thought leadership and sound advice is more useful than any paid-for promotion or advertisement can ever be.

A personal perspective

I ran my own successful residential PR agency before being bought out and brought in by Savills (together with my team of nine) to lead its in-house B2C PR team. This was a brave move by Savills, as not only was it April 2008, but they also had an existing PR function made up of in-house and agencies. They understood that if they were going to excel in PR, they needed to up their game.

Having experienced the best of both worlds, in-house and agency, I always recommend that a company handles its own PR. It is the surest way to get under the skin of a business and to reflect its ethos and culture.

By their very nature, in-house teams are exclusive advocates, in it for the long haul, sowing seeds that may take time to nurture and grow. A very good PR company can be the next best thing, but these are very difficult to find. The best way to find out how good they are is to ask the key journalists who they rate and why. It's in their interest for you to use good people. Mediocre PR representation, whether it's in-house or agency, does you almost more damage than good.

What makes a good residential PR? Some will say the answer is skilful writing, but to me the most important attributes are good manners, common sense and a strong dose of intuition. I learned my craft by reading around the subject and by talking to journalists and asking what they actually needed. Combine that with 30 years of experience, and this is what I've learned about estate agency and PR.

What's the point?

That's a question everyone who considers embarking on property PR should be asking themselves: why bother? After all, estate agents already spend their working lives successfully marketing property. They deal with brochures, advertising and listings day in, day out. So, they may see PR as a fluffy add-on, nice to have but not strictly necessary. They may hate it because, unlike marketing, they have no control over what the end product looks like. Or worse, they may think of editorial as 'free advertising' and expect to have that same level of control.

A significant part of the PR role in estate agency is about educating everyone, constantly, about the positive benefits that it can bring both to the company and to its clients.

The whole point is to create awareness and to achieve specific business objectives, be it through a double page spread about a client's property or through expert commentary on trends, the latest legislation, or what's new in the Budget and what that means for the housing market. The industry is split into B2C and B2B services, which means that there are different audiences to reach, different media outlets to approach and very different tones of voice to convey. For this reason, multidiscipline firms ideally have separate, dedicated PR teams with specific skill sets working closely together but focusing on their own sector.

Residential PR is often the flagship for a brand because it is so visible, its scope so enormous. We all want somewhere to live. Therefore, interest in the subject is vast and ranges from leafing through a glossy magazine, admiring inspirational images of style and design, to devouring five-year house price forecasts in the broadsheets.

B2B PR focusing on the complexities of residential development rather than sales, for example, typically takes place in the trade press. This is an area which some are tempted to avoid because you are effectively talking to your competitors. That's true, but it is vital that you are respected by the competition, and you will also be reaching potential employees. It's all part of building the right profile.

The ability to engage in meaningful, helpful conversations via the media is of incalculable value. Large firms fortunate enough to have highly respected research teams, able to analyse the effects of taxation changes, for example, or rates of housing delivery and the broader economic and political landscape, use this expertise in a long-term strategy to underpin their authority. For those smaller companies without these skills, it would pay to commission tailored research, which they can use to build relationships. Find a gap and fill it.

Because housing is such an important issue both socially and politically, anything that influences market sentiment is potentially a hot topic. There may be a temptation to fire off a comment in a bid for column inches while this or that announcement is making headline news, but knee-jerk reactions should be avoided. Offering considered, balanced opinion is key, and this has to be managed. A firm is, after all, a collection of individuals who may well have different views on the situation, but a business must know its mind and speak with one, clear, well-informed voice, ever conscious of the wider implications of its words.

Media experiences, good and bad

Some people 'get' PR and are brilliant at it. Others don't and aren't. It may be that your CEO or most senior agents are highly experienced and skilled at selling and letting houses, but they may not be the best or most willing spokespeople. If that's the case, find someone else. A weird thing can happen when you put an expert in front of a journalist. They don't always say what they mean to. My advice to agents is always to think of a journalist as the best friend of your best client. That way, the information given is likely to remain helpful and measured. And, of course, I tell them never to say anything they wouldn't want to appear in print or to speak 'off the record'.

I had a client once who was new to PR. He was delighted when the *Financial Times* contacted him rather than going through my PR company. The piece duly appeared, and in the opening paragraph (the one everyone reads), which talked about valuations and asking prices, he was quoted as saying 'The truth is, all clients are stupid.' He rang me in a terrible state, furious that this had been printed, as he had received emails and calls from clients who hadn't appreciated his opinion. As I'd not been involved, I asked if he'd really said it, and he replied: 'I might have done.' I told him it was a gift of a quote and if I'd been the journalist I'd have printed it too. He never spoke to a journalist again without talking to us first.

There's a reason why good journalists go direct, and it's normally when they hope to bypass the gatekeeper and get something controversial. The conversation will be taken down in shorthand and is thus a matter of record. An interviewee may not remember exactly what was said, but the journalist will.

Sometimes PR results directly in a sale, which is fantastic (we once had a picture of a property appear in a national newspaper, and someone who wasn't even looking to move fell in love with it and bought it within a week), but it would be wrong to use these tangible results as the only benchmark. You must take a long-term view, appreciating that the aim is always to raise awareness and promote the right image of your company and profile. When I'm talking at our induction events to people who might not know anything about PR, I mention a few very famous individuals or brands and ask if they have an opinion of them. I explain that this opinion is most likely to have been formed by what they've read about them or heard about them – this is PR.

Bluntly, the more people know who you are and what you do, the more traffic flows to the website, the more potential buyers, sellers, landlords and tenants pick up the phone or call into the office, the more the business grows. But be careful. One of the biggest mistakes in PR is to confuse quantity with quality and to believe in the old saying that all publicity is good publicity. We want to create awareness, yes, but that doesn't mean endlessly pumping out press releases in the hope of picking up a few lines of coverage here, there and everywhere. A brand is known by the company it keeps, and it is of far greater value to have two carefully targeted pieces in publications, consumer or trade, that are right for your clients than 20 mentions elsewhere.

This means that the art of PR can be as much about knowing when to shut up. It isn't necessary to accept every media request and to get involved in every story, but it takes confidence and a real understanding of what the business is trying to achieve to say no (politely, of course).

The one time, however, when it is very important for the PR team not to stay quiet is in a crisis. Every company will have one, if not several, and the worst thing to do is to put your fingers in your ears and hope it will go away. It won't, and as time goes on the public won't remember the incident itself; they will only remember how badly it was handled. So, talk to everyone involved and find out exactly what happened. Is it really a crisis or a storm in a teacup? Acknowledge the situation, respond quickly, apologise (wholeheartedly if necessary; no 'if we have offended anyone' type statements) and explain how the company is going to put it right. Be prepared, be sensible, be authentic.

It's important to remember, too, that the PR team must understand and be mindful of the raft of legislation that estate agents are bound by, not least over anything that could mislead. A good PR may be able to spot a mistake before it becomes an issue.

The media landscape

There's much talk of the changing media landscape, and certainly property journalism was a vastly different place 30 years ago. The scope then for residential PR was very limited, with perhaps just half a page in a handful of nationals. Consequently, editorial coverage was arguably even more highly prized. Then came the rise and rise of the property supplement and a plethora of property-related TV shows. Property was the new cookery. And now, of course, we are seeing countless digital platforms and shrinking newsrooms. In recent years, it seems that well-crafted advertorials are finding an effective place alongside true editorial, perhaps because the way in which we consume the news is also changing. But plus ça change. I firmly believe that although the arena may be different, the fundamentals of good PR remain the same.

From the national and international press to regional titles and online galleries (a gift for gorgeous houses with sumptuous images), there is a strong appetite for property stories, but generating coverage is certainly not a given. For a start, the house, new homes development or nugget of news you are pitching has to be genuinely interesting. It's a bit like other people's children, who, let's face it, often aren't quite as fascinating and talented as their parents like to think. On its own a nice house does not a story make. PRs have to be able to explain that tactfully to the agent, who is obviously trying his or her best for the client. Likewise, that agent needs to know never to overpromise or give any guarantees. Managing expectations is vital. Sometimes an 'average' story goes global while one that looks better in theory fails to land.

The key to making the most of the opportunities is to do your research. Read the property press and websites assiduously. Look at who's writing and what they are writing about. What's the tone? Is it a good fit? Be ruthlessly self-critical in terms of what you are proposing to place and, above all, be realistic. Understand the journalists' job. They aren't there to do you a favour; they are there to inform and sometimes entertain their readers and care deeply about their work. How can you help them do that? If you do decide to get in touch with a journalist to introduce yourself and your company, think carefully first. I know a national editor of a Saturday property supplement who was asked on which day the property pages appear.

As we all know, publicity can be a double-edged sword. While persuading a journalist to visit a house is counted a great success, agents and owners need to be aware that the resulting story may not always be entirely positive. Writers are utterly within their rights to say if there's something about the house they don't like. If that's going to be a problem, then PR is probably not the way to go. There's also the not insignificant consideration that, despite everyone's best efforts, a journalist can sometimes make a simple mistake, leading to incorrect information about a sale or letting.

A truly stunning house and beautiful grounds may well pique a journalist's interest, but add a touch of celebrity stardust to the mix and the story can take on a life of its own. It is absolutely essential to establish from the outset whether a famous client wants publicity and, if so, to what extent they want to be involved. If they are adamant they want no PR, then that's that. We watermark the images of properties on the internet if we think there is a chance they could be stripped and used by the tabloids without permission.

But it is still important to remind publicity-shy clients how the media works. There's a strong possibility that the press will find out if a high-profile person is planning to move. It's very easy for neighbourhood gossip to be picked up by a local paper or news agency, and then the genie is out of the bottle. Once that happens, the result may be a story that runs and runs, and because the agent, client and PR team have had no input, there is a strong risk of inaccuracy. That's an opportunity lost, and it would have been far better to have started out with a well-planned campaign rather than trying to act when the horse has bolted.

Timing and content

We've already looked at why an estate agent shouldn't rush to jump on the latest bandwagon to gain publicity at any cost, but that doesn't mean that PR should ever be an afterthought.

If one were to imagine the timeline of a house sale from the moment an agent is instructed to completion, for example, where does PR fit in? Certainly not as a last resort when advertising has failed to bring forth many viewings, let alone a buyer, and the client is threatening to disinstruct. PRs have many tools at their disposal but a magic wand isn't one of them. In any case, to be brought in after a property has gone on the website compromises the chances of interesting a journalist, who may well only feature it as an exclusive.

Editorial exclusives can be complicated to plan. Advertising, brochures and web listings all have to be coordinated too. It's not unusual for a story to take weeks, if not months, to appear, and so it may be necessary to set a publication deadline on an exclusive. The requirements of both national and regional media also have to be balanced and taken into account. If the property belongs to a household name, for example, the nationals will undoubtedly want the story first, but the owner will be an important figure in the local community too.

It may seem obvious, but before pitching an exclusive, make sure it really is. Check for references to the property online to see what, if any, coverage it has had in the past. This is especially important if it has been on the market with another agent. An internet trawl may also flag up any negative stories associated with the house, which may influence how you decide to proceed.

The PR team needs to be involved from the very beginning. Good advice may help win an important pitch, and every useful piece of information about the property, its history and its owners needs to be discussed as early in the process as possible to give the best chance of publicity.

So, what kind of thing should be music to the property PR's ears?

As already mentioned, nothing beats a celebrity story, just as long as everyone is properly on board, and well-informed commentary on the market is essential. Other than that, it's all about what makes a house or apartment interesting. It could be to do with heritage, who built it, who once lived there or who lives there now. Or it could be because it features the very latest technology or is the third of its kind you've been asked about in a week. Is that the beginning of a trend? A useful indicator is that if you mention it to your colleagues when you get back to the office, it may well have legs.

Case studies are like gold dust. People who swap city life for the country, country life for the city, downsizers, upsizers, families for whom mutigenerational living has or hasn't worked out ... journalists love them, and if you have a client who is really willing to join you on the PR journey then you are in luck.

A joined-up approach

PR has the ability to bring real value and personality to a brand, but, as a key part of the marketing mix, it should not work in isolation. As the gate-keepers of reputation, the PR team should have an important role to play as a sounding board in all communications, watching out to ensure that values on diversity, inclusion, and social and environmental responsibility are properly reflected.

The most effective campaigns are those that bring different strands together to reinforce a common theme. For example, here at Savills the PR, blog and marketing teams worked with our sales and lettings agents on *Voices of Experience*.[1] It was recognised that in times of uncertainty, making decisions about such an important asset as a house can be very difficult, and sound advice is, therefore, all the more sought after. An advertising campaign was launched with the strapline *Now more than ever, our 150 years of experience matters*. As the advertisements appeared, so too did a series of blogs on the Savills website in which experienced residential agents recalled facing market uncertainty in the past, the lessons they learned and the expertise they were able to pass on. There then followed a social video campaign across all our social channels, including Facebook, Twitter, Instagram and LinkedIn. It goes without saying that the experience theme was also explored in the press.

Creative thinking

Competition for coverage is incredibly fierce, and creative thinking helps you stand out from the crowd. Be bold but not reckless is my motto, because never forget that in estate agency we are taking care of not only the company's reputation but also our clients' property interests. Property is a personal business, and someone's home needs to be handled carefully and with sensitivity.

Savills' 2019 campaign, the *#JarfromAfar* challenge,[2] is a case in point. After 24 happy years in their Scottish farmhouse, the decision to sell and downsize was not an easy one for our clients. When the photographer arrived to take pictures for the brochure, the vendor was tearful about the impending move. So, having discovered a shared love of Marmite, the photographer cheered her up by hiding a jar in every shot, challenging her to find them. The story could have ended there but for a member of the PR team realising the potential. We challenged readers of our blog to find the pots, and the story appeared on social media and online from *Estate Agent Today* to *The Sun*. It even made it as far as Australia.

Conclusion

Whether you are reading this as an estate agent keen to understand how PR will help your business or a PR professional eager to carve out a career in the residential sector, the most important thing to remember is the significance of reputation. PR can be the single most important, cost-effective way of building and protecting that reputation, positioning you as well informed, approachable and trusted in an industry that is not always regarded as such. But remember too that a good reputation has to be earned and rooted firmly in fact. Meaningful PR takes commitment and time; it is not simply about being in the public eye or spouting endless soundbites. You can't dip in and dip out. It's not a quick fix but a long-term investment, and done properly, will pay dividends.

Notes

1 www.savills.co.uk/insight-and-opinion/tagged-articles/Voices-of-Experience [Accessed 14 October 2019].
2 www.savills.co.uk/blog/article/282060/residential-property/take-the-jarfromafar-challenge-and-have-a-look-inside-harperrig-house.aspx [Accessed 14 October 2019].

11 Promoting B2C proptech

The role of PR in building a proptech business in estate agency

Ed Mead

The need for change

Look back into the mists of time and try to imagine what buying, or renting, a property was like. I would surmise that it wasn't really very different from today: esoteric, long winded, managed by people you wouldn't choose to spend quality time with, and with a lot of people to pay – whether you actually bought the property or not.

The fact is that the basics really haven't altered for generations – certainly in the almost two generations that I've been doing this. We all know how flaky the general property buying process is, because those suffering at the hands of the process, disgruntled by the expensive failure of a purchase or sale, talk to someone – often a journalist, who will then let their readers know. PR and the media have been an integral part of informing the (lack of) process to the buying public for many a year. How else would gazumping have become common currency?

For many, the negativity and stress inherent in buying and selling are in direct proportion to the length of time it takes to complete the process. Indeed, over the last ten years alone, it now takes 50 per cent longer to complete a purchase. According to Matterport's 2019 survey of 2,000 buyers and sellers,[1] the average time from offer being agreed to exchange, that is, when a property is actually sold, is now over five months – more than twice as long as it was when I started in the industry in 1979.

So why has the process stalled, and why isn't negative publicity fuelling change?

It's partly because we're British, and we don't complain. It's partly because estate agents have no united voice, or regulation, or barriers to entry – or sanctions at all, in fact. And partly because no one can afford to tell the buying public that there are the tools to make their lives easier. Yes, there's an over-worked Property Ombudsman, but I would venture to suggest that the biggest problem is estate agents themselves. Don't get me wrong; there are plenty of good ones out there, but as usual, the bad ones attract brickbats and give the majority elsewhere a headache.

The arrival of the online agent

The buying and selling ecosystem is being disrupted as never before, since April 2014, by the only arrival in the property space willing to spend countless tens of millions of pounds telling the public, directly via their TV screens, that there is another way of selling a home. Whatever your view (and high street agents have an unremittingly negative one) of Purplebricks, you don't go from zero to the biggest agent in the UK in five years without doing something right – or at least appealing to something in the public's psyche.

Interestingly, back in the late 1990s before portals like Rightmove and Zoopla existed, various focus groups were set up by the Primelocation management team to see how the potential buying public might react to an aggregator – a portal – capable of showing them many properties from different agents on one website. The focus groups had to be abandoned because the public misunderstood what a portal was and were over-excited by the prospect of not having to deal with an agent at all (in fact, they spent the sessions complaining about estate agents).

The odd thing about Purplebricks is that while they got their name recognised quickly, potentially saving a fortune on their own PR and marketing budget, so many agents have assiduously sought, through PR and other means, especially social media, to denigrate them. Is there any truth in the old adage 'All publicity is good publicity?'

A missed opportunity?

Property technology – proptech – didn't really start to have the ability to impinge on the buying process until three or four years ago. Until then, the only things the public might have noticed that improved their experience were photographs, portals with clunky interfaces, floor plans and email. Improvements had come at a snail's pace.

But today there are some fabulous add-ons in the areas of communications, information and the physical ability to see properties when you want, virtually or in the flesh. The problem is that most – or *all* – proptech companies simply don't have the budget to promote their product to the general buying public.

The launch of Viewber

Take Viewber, the company I left my estate agency job for to start in late 2016. I recognised that buyers mostly wanted to see properties at weekends or evenings – times when most estate agents are closed. For many years, buyers had grudgingly accepted this. But as millennials enter the property market, there's no way they'll want to be told that they can't see a property when they want.

At my previous employer, Douglas and Gordon, our weekend diaries used to be full by Tuesday, but a huge number of potential weekend viewings were missed because our staff, like most, didn't want to work weekends. So, I surmised that having a flexible UK-wide army of people willing to open doors out of hours when booked by estate agents online made a lot of sense. Unfortunately I didn't have a massive PR and marketing budget available to tell anyone thinking of buying a property that there was now a method whereby they could see it when they liked, cause them all to rush to their agents complaining when refused weekend viewings, and thereby force agents to start using the service. But I capitalised on other promotional opportunities.

Box 11.1 Viewber

In 2016, Ed Mead surveyed 250 UK buyers and tenants of residential properties regarding their opinions on viewings:

- 97 per cent said they would find it beneficial to be shown around a property at evenings and weekends.
- 62 per cent believed that they had missed an opportunity to buy or let a property because they weren't able to view it at a suitable time.
- 81 per cent would be willing to be shown around a property by an independent vetted person.
- 62 per cent would feel more comfortable being shown around a property by someone who wasn't looking for a sales commission.
- 82 per cent would expect to be able to arrange a property viewing at a time to suit them.

These interesting results pointed towards a new form of estate agency, which makes out-of-hours views possible and capitalises upon online communications without sacrificing the vital face-to-face contact necessary in the sales process. In September 2016, it led to Ed Mead and the entrepreneur Marcus de Ferranti founding Viewber. Financial backing from former Foxtons and Marsh and Parsons boss Peter Rollings followed quickly.

By March 2017, the company had 1,000 registered Viewbers (viewing assistants). During the same year an average of 30 new Viewbers signed up every day, and approximately 8,500 viewings are now carried out each year. Viewber has over 500 registered clients, ranging from lettings, sales, auction and commercial agents to property and asset managers.

Viewbers are located all around the UK. They are vetted and hand-chosen for their professional approach, and being local, they know the area well. Most have worked in the property industry previously.

Viewber's USP is 'any viewing, any where, any day, any time'. It provides staff to host viewings and open house events; viewing assistants for a defined period or tour; inventories, inspection reports, photography and 360-degree tours.

In February 2017, *The Sunday Times* listed 226 occupations that were ripe for automation. Third on that list, with a 97.29 per cent chance of automation, was 'Real Estate Broker'.

Proptech will undoubtedly have a huge impact on the way people buy, sell, rent and invest in property. From artificial intelligence to apps, it should become quicker and easier to find the perfect property and move in. As well as making life easier for the public – quicker searches, more relevant marketing and digital conveyancing – proptech will revolutionise estate agency: improving efficiencies, bettering customer service and bringing the industry in line with other service sectors.

Viewber is making a hybrid agency a reality. For estate agents planning to streamline physical branches in favour of a virtual operation, viewings over distance can become an issue. Viewber allows agents to centralise with fewer staff but still offer accompanied viewings anywhere in the country.

As such, Viewber challenges the idea that proptech is disruption: for Ed Mead and his 'Viewber army', proptech is all about collaboration.

Getting the message out

Over the years my PR profile had risen, a corollary of the fact that the business I worked for at the time, Douglas and Gordon, employed a PR firm, Jago Dean, who taught me everything I now know. For ten years I wrote a column as the 'Agent Provocateur' in *The Daily Telegraph* and over the same period also wrote weekly as the *Sunday Times* Property Expert with multiple pundit appearances and two BBC property shows. This obviously helped.

Because of this huge PR profile in the property industry, leaving my job was big news and enabled me to parlay some of what was going to be happening with Viewber into many column inches, and my phone rang off the hook with agents wondering what Viewber was. This would have cost hundreds of thousands of pounds in marketing. The next stage was to persuade a bona fide journalist that what I was doing was newsworthy. I thought it was, and luckily the well-respected Jessie Hewitson agreed and did a lovely piece in *The Times*[2] about how what I was doing might affect those looking for property. The results were equally staggering, with more than 500 'Viewbers' – what we call our viewing agents – signing up in the week the article came out.

The momentum gained from the articles in *The Times*, *The Sun*, *The Mail*, *The Negotiator*, *The London Magazine*, www.propertyindustryeye.com and www.estateagenttoday.co.uk was unprecedented and enabled the company to grow in a third of the time a standard marketing campaign would have done – and it was free.

That was all well and good, of course, but the national coverage was merely the start of the journey of trying to tell buyers that agents do have a choice in how to deal with their out-of-hours requests. Such campaigns have to be backed up, and as I did not have a vast B2C budget to enable me to put my proposition in front of the nation's buyers, I continued to drip-feed news into the national press. Importantly, news *has* to be of interest to readers, and journalists have a nose for what that might be. It's what makes them good journalists and why a PR company's job is so difficult.

Viewber had an extra hurdle to jump insofar as it was a brand-new idea that could have been perceived as stepping on agents' toes. In a notoriously conservative industry, getting across the message that the aim was to add to what they were already doing was tough. Trying to imply that agents could reduce staff and still cover ground was a tough one – and, of course, we had a number of different and potentially clashing customer bases. High Street agents – an obvious early client base – aren't fans of online agents, and private landlords love Viewber, possibly at the expense of using letting agents. We thought all could benefit, but messaging is always difficult to nuance. At first, we tried to be all things to all potential clients, and as it became obvious which bases we were covering, we tried to narrow it down. A B2B marketing budget doesn't leave much leeway, though, so picking your battles is necessary. Interestingly, some of our early customers didn't want us to use testimonials – partly as I think they were embarrassed to be seen using us (no idea why), but also they didn't want their competitors to gain the same advantage and start using us.

Successful media relations

So, next was to keep the brand in people's line of sight. Interestingly, we never really had a fixed PR plan but felt we'd offer the right journalist exclusive access when the right angle came along – so getting to 50K viewings, each annual anniversary or a fundraising milestone all made for a potential story. A good relationship with Myra Butterworth of the Mailonline, allied to a good lunch, persuaded her that it would be fun to be a Viewber for a morning and report on the outcome. Clearly the strategy was fraught with potential downsides, but when you're trying to grow a business from ground level, particularly in an area where your intellectual property is hardly unique and the capacity for copycats is substantial, speed and brand recognition are important, so you have to take risks. In 2017 Myra spent a morning as a Viewber and showed a couple of flats on behalf of estate agents. The resulting coverage was huge and, with the Mailonline's reach, had a huge impact on our brand recognition with both estate agents and potential 'Viewbers'.[3]

It's a funny thing, PR. For many it's simply about churning out press releases, and that can play a part. Social media, and particularly Twitter, has made the hunt for column inches a touch more meritocratic, but there is serious value in keeping one's powder dry. Some choose to be ubiquitous, but their currency is diminished as a result. Most respected journalists, working in the national arena with their own reputations to think about, appreciate a considered and thoughtful approach that gives them pretty much all they need, wrapped up in something they know their readers will understand. For many years I saw the merry-go-round of property correspondents as a negative, merely a tick on the journey to an editorship and including economics, motoring, sport plus a few others. But a characteristic of the kind of journalist you *want* to be writing about you, or your business, is that they know their readers. Exactly what they're writing about might be considered secondary, as they'll get it right whatever it is.

In the quest to find potential clients, the property trade press was important but limited. There may be a plethora of websites and publications that report weekly or monthly, but their impact on those reading them varies wildly. Social media, and the 24/7 nature of news, means that the most dynamic sites have to produce stories daily. This is good because of the sheer volume of stories needed to fill the site, and bad because your limelight will last 24 hours if you're lucky. In the case of the residential property sales world, three titans dominate the arena: *Property Industry Eye, Estate Agent Today* and *The Negotiator*. All are curated by well-known property journalists – Rosalind Renshaw, Graham Norwood and Nigel Lewis, respectively. All know instinctively when they're being sold to, but again, many years of painstaking grooming of your own PR profile will pay dividends in the long run. In my case, Ros asked me to write a column once I left Douglas and Gordon, and of course that gave me a platform, but one to be used sparingly. Writing as the 'Agent Provocateur' in the *Daily Telegraph* property section for ten years, I aimed to be as objective as possible, but I did find the brief more difficult as I didn't want to put off potential customers, which I might have done had I been 100% honest. Being perceived as an 'expert' at anything is a useful moniker, and the property market is no different.

Knowing what constitutes something worth trying to get coverage on will vary depending on the outlet you're targeting. Social media has to come into that today and, of course, is an excellent way to build a profile as an expert with a human face. But very few stories start on Twitter.

Clearly, column inches in the trade press directed specifically at your target, in my case estate agents, will – for some of the reasons mentioned earlier – draw negative comments, often anonymous, hurtful and self-serving, below an article. But the subject matter still has to be of interest to the readership. In my case, reporting numbers of viewings done, simply surviving another year, partnerships with well-known agents and reports from workers in the field were often considered newsworthy and have been regularly reported in *Property Industry Eye, Estate Agent Today* and *The Negotiator*. Looking up mentions of Viewber on any of those sites will show how effective it's been in establishing the brand.

For many companies, and proptech or online estate agency is no different, cultivating a relationship with a journalist is helpful, and even better with a respected national publication. In my case, further articles as the business grew were forthcoming, with follow-ups from David Byers[4] and Jessie Hewitson[5] in *The Times*.

Many journalists didn't bite on stories associated with Viewber, but there are multiple ways of influencing the public. These include social media, and I have found that Twitter gives a good opportunity to build a personal brand that people can trust – and ultimately all tweets point back to the Viewber website. It's given me many thousands of followers and access to both journalists, commentators and potential clients – well worth some effort. I always posted myself and never outsourced this function; it's intensely personal.

Earlier, I mentioned that being an expert and continuing to keep your finger on the pulse enables those journalists with whom you have a relationship to continue to canvas your views on the property market. These comments are always attributed, in my case, as a property expert and founder of Viewber. Press comment has continued, as have regular visits to the stage for opportunities including the FT Festival, op-eds in the FT, the RICS residential conference and regular slots on LBC and Talk Radio, despite my not having been on the front line for almost three years. Subjects discussed are typically topical, Q&A based, and relate to housing supply and demographics as well as more considered areas such as financing, performance of large estate agents and online agency. Nothing seems to beat experience and a bit of confidence in front of the camera or microphone. Back in the 1990s and 2000s I fronted two TV series for the BBC, *Chelsea Tales* and *Under Offer: Estate Agents on the Job*, and they probably helped a lot.

Finding someone within your organisation who will come across to the public as trustworthy and engaging is not easy. A very well-known agency represented throughout Britain and employing thousands had, as far as I can tell, only one person safe to be put on screen. It's more about delivery than content, so if you're looking for someone to be the PR spearhead it might not always be the CEO or MD – think out of the box.

Putting a value on expertise

Knowing and believing in PR doesn't make it easy to apportion monetary value to it. The reason a property owner decided to use you to sell their house is unlikely to be given as 'I saw your company on TV last week', but brand recognition will most likely be a strong party to that decision. It's a difficult process to be objective about. What will help is keeping blog content fresh so that if anyone looks the last post won't be from months ago.

It's safe to say, though, that the brand recognition gained by Douglas and Gordon meant we always punched well above our weight. With Viewber, the audience, both public and agents, meant that we managed to get our business the sort of profile that would have cost many hundreds of thousands of pounds to achieve in pure marketing alone. Many will simply put their PR spend down as marketing, but it's not marketing and requires a very different mindset. Needless to say, I think it's a vital part of any small proptech company's growth.

Notes

1 www.estateagenttoday.co.uk/breaking-news/2019/5/revealed--exactly-how-long-it-takes-to-buy-a-home-in-the-uk [Accessed 15 September 2019].
2 www.thetimes.co.uk/article/what-it-really-costs-to-sell-a-property-dv25gl05d [Accessed 15 September 2019].
3 www.dailymail.co.uk/property/article-4500170/Viewber-earn-extra-cash-showing-potential-buyers.html [Accessed 15 September 2019].
4 www.thetimes.co.uk/article/show-homes-the-job-that-pays-you-to-nose-around-c7ltn87r3 [Accessed 15 September 2019].
5 www.thetimes.co.uk/article/how-technology-is-changing-homebuying-ng2p8smf2 [Accessed 15 September 2019].

12 Promoting B2B proptech

An online revolution in the private rental sector

Louise Parr

Introduction

That proptech is rapidly transforming the property market is undisputed: the 'disruptive' force of technological change is driving advancements at a rapid rate. But look beneath the surface, and the transformative effect is far from universal: many subsectors in property are in fact reluctant to adopt new technology, and of those who are doing so, few are realising its true potential as a communications tool.

The challenge for start-up organisations in proptech is that it is a noisy market with many diverse and innovative services. From a communications perspective, being clear on who you are targeting and how your solution will help them has never been more important, along with attendance at the right events with the right contacts.

This chapter will explore both proptech's potential as a communications tool and the opportunities for property PR and communications professionals to boost the take-up of proptech, while also looking at the challenges in doing so.

Some consider proptech to be just one part of the digital transformation of the real estate industry. Others see it as what makes the real estate industry more dynamic and enables it to innovate further. But irrespective of the definition of proptech, it is an *enabler*.

Spanning both the public and private sectors, the property sector is extremely broad, which makes the platforms for PR both wide ranging and varied in terms of target audience, the channels used and the choice of language. Previously, *the public sector* and *digital transformation* were rarely found in the same sentence, but this is changing rapidly, and PR is integral to the challenge of boosting take-up.

This chapter is based on the experiences from Kamma (formerly GetRentr) following the company's rebrand in September 2019. The chapter gives an insight into the way in which Kamma is helping to shape communication in the industry and the communications challenges faced. It also demonstrates the importance of partnerships.

Box 12.1 GetRentr: the story so far

We had been through quite a journey as GetRentr: we started as a rental app, but our path has meandered through myriad unforeseen territories and we are now much more. We were founded on the core principle that 'Nobody should have to live in unsafe or unsuitable accommodation ever again,' and even through our journey has wandered, our mission endures.

When we started three years ago (2016) as a service to connect tenants, landlords and tradesmen to improve the rental market for all stakeholders, we had visions of topping the app store and being on smartphones across the UK. As we began to research the market in depth and build the product, compliance became a key pillar of our offering, as we wanted to make it as easy as possible for landlords to stay on the right side of the law. One aspect of compliance that we realised had to be part of this offering was property licensing: compliance with the dynamic, diverse and difficult set of geographical regulations governing minimum property standards in certain sections of the private rented sector (PRS).

As we started to gather the data, we quickly realised why nobody had fixed this problem: it's a total minefield. We spoke to the industry and understood that, while there were a number of innovative young companies looking to streamline the property management function, nobody was looking to solve property licensing. Meanwhile, the PRS was rising up the political priority list and enforcement was really beginning to ramp up: there was a genuine desire for an automated user-friendly platform to advise licensing requirements on a property by property basis. So that's what we built. ...

Having honed our data gathering and aggregation techniques to crack the problem of property licensing, we were looking for the next challenge. What had become clear on our journey so far was that building the product with a slick, simple interface was the easy bit. The reason why nobody had been able to build a property licensing solution before was the data side: finding, tracking, gathering, cleaning and reconciling disparate messy data was really the problem that we had solved. If we could do it with property licensing, why not other datasets based on geographical regulation of UK property?

This confirmed what we already knew: we weren't a rental app anymore. We had morphed into a company that harnessed data to answer difficult business questions at scale. We were combining multiple datasets to drive unique and powerful insights into the property market for multiple stakeholders. While we were born as GetRentr, we have grown into something else, and this name no longer reflected our offering.

In the search for a new name, we started looking at translations in every language imaginable for data, intelligence and anything else we could think of, but nothing captured our imaginations. We took a step back and thought about what it is that we fundamentally provide to our clients. We realised that the clever technology, the intelligent data manipulation and the elegant front-ends were all secondary to what clients rely on us for.

Our clients depend on us for the clear and precise results of complex analysis. The algorithms are not important: nobody decides to use our platform by looking at our codebase. Clients decide to use us by looking at the results that we provide, and this is something that we never want to lose sight of. In order to continue and build upon our success, our focus must remain not on methodology but fixated on the results we give to our clients.

Given that results are fundamentally what we do as a company, we thought: 'Let's just call ourselves what we do.' Kamma is a Sanskrit word meaning 'Results'. Through our data and insight, we provide answers to our wide range of clients, in detail and at scale. Our products are built to layer clean, user-friendly interfaces across the complex machinery of our data gathering and cleansing. Through clever use of technology, we convert disparate, messy data into precise, actionable results, driving business value.

At the time of writing, GetRentr was in the process of rebranding to Kamma.

Using content to create opportunity

'Education and knowledge of the 400 plus rules and regulations on lettings, including licensing, is poor in this country. Lenders, insurance and legal companies don't insist on any evidence a property is being let legally, including sight of any required licences prior to supporting a property's purchase so if a landlord doesn't use a qualified ARLA Propertymark, NALS, UKALA or RICS agent who are most likely to be trained in lettings legals, properties can be bought and rented without the landlord having a clue of their responsibilities.'

Kate Faulkner, property market analyst and commentator and one of the UK's top property experts, describes the current communications challenge[1]

Box 12.2 Changes in the letting sector and Kamma's proptech solutions

The PRS has grown substantially,[2] and the Ministry of Housing, Communities and Local Government (MHCLG) and local authorities are collaborating to improve living conditions for renters now more than ever. From tackling rogue landlords and establishing agent databases to introducing stricter controls on health and safety, these have significant impacts for UK letting agents.

Tighter regulation of licensing laws applicable to HMOs

The PRS is undergoing significant change, particularly in relation to houses in multiple occupation (HMOs), for the better protection of tenants.

HMOs form a vital part of this sector, often providing flexible accommodation for less affluent tenants. HMOs are commonly known to be occupied by students, but there is also a growing number of young professionals and migrant workers sharing houses and flats. In 2018 MHCLG issued new guidance for local authorities to implement changes to HMOs following a series of new guidelines to help improve living standards.[3]

Property licensing exists to ensure that residential accommodation within the PRS is safe, well managed and of good quality, with a particular focus on safety. It can be complex for letting agents to know what rules and regulations apply to properties on their books.

In October 2018, the government mandated some much-needed controls regarding HMOs[4] and the process of turning a single building into multiple dwellings. When rental circumstances change, new conditions and licences may apply. Sometimes this is missed by the landlord as an honest mistake.

In April 2018, the MHCLG published statutory guidance for a database of rogue landlords and property agents under the Housing and Planning Act 2016.[5] This aims to name and shame any landlords and agents who have been prosecuted so that tenants can make informed decisions about the properties they rent.

Kamma's unique product displays a dashboard of properties within a portfolio, highlighting the aggregate of which properties require mandatory, additional and selective licences. Users can then drill down into individual properties to analyse in detail their licensing requirements. This ultimately enables users to avoid fines and protect their reputations.

The Homes (Fitness for Human Habitation) Act

An update to the Landlord and Tenant Act 1985, the Homes (Fitness for Human Habitation) Act 2018[6] is having a positive impact on the collaboration between local authorities, tenants, landlords and agents. The Act will provide local authorities with the power to enforce new licensing needs to ensure better property management and tenant protection within their boroughs.

Managers/landlords of HMOs are required by law to ensure that the electrical installation is regularly tested and up to date, with certificates to prove this when required by the local authority. This is another good step in relation to the protection of tenants by local authority regulation enforcement but a further obligation for agents and landlords to manage.

Furthermore, additional licensing requirements may apply to specific properties and addresses – meaning that there are additional conditions and often updates needed prior to a property being officially granted a licence for tenancy occupation.

Kamma helps in providing landlords/letting agents with property licensing requirements, whether mandatory, selective or additional, detailing the dates those licenses expire and the need for renewals.

As Box 12.2 shows, the PRS sector has evolved substantially, resulting in significant changes to business practices for letting agents and landlords in particular. Kamma uses proptech to connect legislation to the built environment, helping multiple stakeholders to understand the requirements and obligations of property licensing, designed to raise minimum housing standards for all in the PRS. Kamma helps local councils, agents, lenders and landlords to identify whether a property needs a licence and whether it has a licence, and constantly monitors whether any changes to the licensing framework affect the property. Not only is this beneficial in protecting vulnerable tenants from exploitation; it also helps organisations mitigate risk and avoid hefty fines.

Hence, Kamma's content marketing and PR plan takes into account many of the recent changes in the PRS, communicating the change at a national and local level to relevant audiences. In aligning the brand with such significant news stories, we can highlight the fact that this issue is real and growing, providing multiple sources. The appetite for this information has been shown to appeal across social media channels, creating new followers and audiences as well as driving traffic to the website, which often converts into product demonstration opportunities for our sales team. Our key predictions for 2019 in the PRS continue to drive traffic to our website eight months after it was first published.[7]

In-house content creation is key when the marketing budget is low. Also, the marketing team's remit is wide ranging, covering a wide variety of communication tactics in addition to content creation, PR and brand building – especially during a rebrand. Building authority with unique content favours the company in Google ranking terms and search engine optimisation (SEO) – in Kamma's case, in relation to property licensing and associated topics and news stories. New legislation, revisions to government Acts and other information contained in Box 12.2, for example, are used to inform a series of articles hosted on the Kamma (then GetRentr) website.

Property licensing is diverse, with many different groups impacted by its legislative obligations. Kamma has taken on the role of 'curating' the diverse views by using social media to relay how each of the affected groups is responding.

Kamma also provides its substantial audience with links to relevant news stories. For example, we might use a breaking news story featured in *The Guardian* and convey snippets of the story to our audiences via Twitter. This provides a free service to our target audience, while also benefiting media relations by raising awareness of Kamma among journalists. Furthermore, by tagging an industry organisation such as ARLA Propertymark or the National Landlord Association on Twitter, Kamma is able to reach a broader selection of UK estate agents.

News stories regarding penalties for letting agents and landlords have become common and frequently appear on local council websites, alongside quotes from relevant councillors describing a 'clamping down' and a drive to better protect vulnerable tenants. The PR opportunity for Kamma is to include content associated with these news stories on our website, which is often picked up by journalists and features on websites relevant to landlords and agents, such as Estate Agent Today, Landlord Today, Property Industry Eye and The Negotiator. This drives traffic to our site and in doing so increases demand for a demonstration of our services. A recent example of this is a September news story roundup we created following a series of articles published in *The Times*[8] following an investigation into the PRS and various online letting sites. This is published on our website, and consequently Kamma will likely be featuring as lead in future articles.

The power of collaboration

Industry events are great for keeping in touch with the market, understanding how circumstances are changing and generating new business leads. Kamma's platform lends itself to product demonstrations and is therefore well suited to events – but the challenge is that people are increasingly time poor.

In May 2019, Kamma attended Future Proptech 2019 (FPT2019), had a shared stand negotiated in the Agent Zone and negotiated a free speaker slot for its CEO, Orla Shields. The presentation focused on the changes in the professional data landscape – how data is now often said to be more valuable than oil in today's society due to the volume consumed on a daily basis, and questions concerning how useful this data is, what can be done with data and how it can be made relevant. The event was a huge success because it provided direct access to our target market, helping to close the communication gap between Kamma and agents.

Many collaboration opportunities with potential partner organisations resulted from the FPT2019 event, creating new and unexpected PR opportunities.

One such angle was a piece of research created by Digital Risks,[9] which considered the gender bias when creating start-up organisations in the technical sector. This provided an opportunity for Kamma marketing and the CEO to respond, stating that women entrepreneurs in the UK had founded over 500 still-active start-ups in the last 18 years. The article not only provided great insight for entrepreneurs but also highlighted the disparity of venture capital investment between genders. As a result of the coverage, Orla Shields and Kamma were positioned among other prominent UK start-ups, interviewed and featured on various channels. Between February and April 2019, Orla Shields took on a prominent role (including in the context of International Women's Day) and featured on key websites including Value Walk, Bitcoin Warrior, TechTalkStart, NexChange Group and Compelo. Kamma's social media team capitalised on the opportunity, retweeting content for maximum exposure. These were all published on Kamma.com[10] and linked to the relevant publications.

Reference to articles and quotes has proved very beneficial in creating links for SEO purposes. Hashtags are also a great way of growing a social media following, and #IWD2019 (the International Women's Day hashtag) worked well in this example. In other circumstances, #TechTuesday has been very useful: #TechTuesday is a great weekly Twitter conversation which allows us to successfully repurpose content (see Figure 12.1).

This steady stream of change has provided several PR opportunities for Kamma. It has allowed us to broaden our connections, to be bold in our

GetRentr @Getrentr · Jul 16   ⌄

It's #techtuesday!

@BitCoinWarrrior syndicated an interview with the @Getrentr CEO,
@Orla_Shields - which originally appeared in @valuewalk .

You can read the full research and interview here 👆

ow.ly/VmNw50v1PZG

#bitcoin #bitcoinwarrior #bitcoinforbusiness

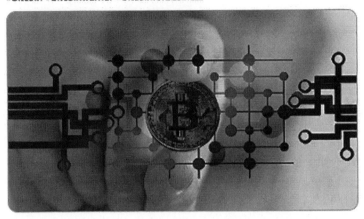

Figure 12.1 #TechTuesday.

social media approach, to tag authoritative content providers and thank
them for their research, and to share this with our followers. The creation
of daily news feeds is time consuming, so repurposing third-party content
across social media is not only efficient but also crucial in building trust and
authority with new audiences. Developing content to support these changes,
we can then refer target audiences to our online channels to find out more
about us. We have seen a steady growth of our social audiences, which is
consistently engaged – something we felt was a good indicator of our value.
This has all been achieved without any marketing budget.

Box 12.3 Kamma's partnership with ARLA Propertymark

The June 2019 updates to the Tenant Fees Act were made to prevent unclear
costs and administrative charges by agents, which had been seen as a form of
tenant exploitation. Membership bodies such as ARLA Propertymark provided services to inform and educate letting agents, via webinars, regional
conferences, seminars and content marketing, advising on how they need to
update their business practices.

In July 2018, ARLA Propertymark partnered with Kamma. A series of marketing initiatives were agreed, including attendance at marketing conferences and emails to agents on licensing regulations, as well as datasets from Kamma to ARLA Propertymark on their website for upcoming schemes and consultations.

The service was promoted with an email to all members from ARLA Propertymark, website updates, and then regular bespoke emails and data insight provided by Kamma.

Figure 12.2 Kamma and ARLA Propertymark's collaboration.

Another huge benefit of any technological advancement is the opportunity for organisations to collaborate and share ideas. Through Geovation,[11] Kamma has worked with both HM Land Registry (HMLR) and Ordnance Survey (OS), not only allowing us to access data to help our platform grow but also benefiting HMLR and OS in working with Kamma to explore novel and disruptive uses of their data.

This in effect means going against the norm and, in turn, helps develop business and datasets through learning what proptech firms today need access to in an appropriate format. This collaboration evolved from Kamma successfully entering the Geovation Accelerator Programme.[12]

HMLR went on to be recognised by the publication *Estates Gazette*, with a Big Picture award recognising its 'collaborative thinking' and 'pulling together of the world or real estate and technology', going to show how a partnership approach like that of Kamma and HMLR can reverberate through the industry, creating recognition and PR opportunities in its wake. Kamma has also featured on the HMLR guest blog, which creates awareness and authority for Kamma as a brand.[13]

Box 12.4 Kamma's involvement with the Geovation Accelerator Programme

The Geovation Accelerator Programme is run by HM Land Registry and Ordnance Survey. Geovation is 'a community of location-data and proptech collaborators looking to make positive change happen'. Since its inception in 2009, Geovation has become a leading proponent of the value of open innovation in the public sector. After opening its first space in summer 2015, Geovation has grown to support a community of more than 1,500 entrepreneurs, investors, developers, academics, students and corporate innovators.

The service targets start-ups in need of location or property data and expertise. The Geovation Accelerator Programme provides up to £20,000 in grant funding, access to data, product development advice, geospatial expertise from Ordnance Survey and land and property insight from HM Land Registry, as well as business mentorship and coaching to help prepare for presenting to investors from the wider team and our partners. It holds community events and hands-on development resources, and advises on business development from initial vision to sustainable growth. Success is measured by the success of the start-ups in receipt of the advice.

As it is an initiative of Ordnance Survey and HM Land Registry, supported access to their datasets is provided. This includes detailed maps and addresses, ownership information and valuations. Making this information available to start-ups helps unlock greater value from these national data assets.

The Geovation Accelerator Programme benefited Kamma by providing free rented office space in the centre of London, a presence among a huge range of GeoTech and proptech experts with connection into organisations and a network to help Kamma accelerate its journey, as well as a £20,000 grant, which helped investment priorities in the platform and some very positive publicity, helping to drive interest in the product.

> 'Kamma is a really good example of just the kind of company we were looking for when we established the proptech stream at Geovation. They're taking our data, and they're using it in really novel and innovative ways that we would never have foreseen ourselves as an organisation.'
>
> Andrew Trigg, chief geospatial and data officer, HM Land Registry

---

How technology is disrupting property processes

The sale of a property on Instagram in January 2019 highlights a possible new trend and proves how valuable free social media channels can be in promoting property, just as with any other product. The property industry, in communications specifically, is increasingly making greater use of social media.

Smartphone apps to promote rooms to let are becoming increasingly popular, and 'dating style apps' are likely to appear on the market imminently, aiming to bring tenants together and allow easier, safer and less expensive ways to find new housemates and to promote vacant rooms. One such app is Barcelona-based Badi, which has proven itself in Spain and, at the time of writing, has just landed in London, where it aims to steal market share from Spareroom, Rightmove and lettings agents.

Blockchain has become something of a buzzword in proptech and is used by many organisations.

Blockchain, according to *Harvard Business Review*, is '[a]n open, distributed ledger that can record transactions between two parties efficiently and in a verifiable and permanent way'.[14]

Clearly, Blockchain can add huge value as a technology to connect all areas to assist property completion – from digitally signing a mortgage, to having the entire mortgage transaction processed digitally, and then digitally registering with HM Land Registry. However, the entire process is still in its infancy. We are yet to see examples of this working in property purchasing, but this is an exciting future opportunity for potential collaboration between mortgage lenders, conveyancers, home buyers and HM Land Registry.

Given the vast range of technology at our disposal, the world of mortgages and the house-buying process in the UK can still feel outdated. Prior to 2018, there were no digital mortgages recorded in the UK. The first digital mortgage was signed on a property in Rotherhithe, East London in April 2018.[15] Since then, the service has been tested in a controlled environment. Coventry Building Society and Enact Conveyancing have worked together to pilot the service with a number of borrowers. In 2019, HM Land Registry confirmed that a number of big name high street banks have signed up to use the service, enabling their customers to sign their mortgage deed online, with hundreds of digital mortgages now being registered.

In April 2019, Nick West, chief strategy officer at Mishcon de Reya, commented:

> It is clear that, as it stands currently, real estate transactions are too slow. This affects buyers and sellers, particularly where transactions fail or are abandoned and at great cost. Technology, and especially blockchain, coupled with improved data management has a huge part to play in revolutionising this process. What's exciting here is that this is a real world application of blockchain technology in our legal market. If we can turn this proof of concept into reality it will be of significant benefit to anyone buying or selling real estate assets in the UK.[16]

At the World Built Environment Forum in New York, May 2019, the RICS Tweeted: '#Blockchain as a technology has the potential to support #realestate transactions. Blockchain has the technology that can create a #ledger of truth, as it will allow systems and process to sync in a simultaneous matter, says Jeffrey Berman, @Camber_Creek.'[17]

In a blog, HM Land Registry demonstrated substantial demand regarding interest in digital mortgages, but stated that the technology that underpins them across the banking sector is not yet there to make this a reality.[18]

HM Land Registry aims to transform the conveyancing market[19] through quicker and simpler digital services and improved use of technology, making transactions instantaneous where possible and simplifying the homebuying process.

However, HM Land Registry has made significant advances regarding the registration of the deed following months of collaboration and testing with Coventry Building Society and Enact Conveyancing and uses GOV.UK Verify[20] to enable borrowers to securely verify their identity before digitally signing their mortgage deed online.

Another exciting use of digital technology in proptech is that it is allowing mundane activities to be automated. The use of augmented reality such as chatbots to process simple frequently asked questions online allows the redeployment of staff, ensuring business efficiency. In the case of letting agents, for example, the use of chatbots can free up time to focus on viewings and property sales, helping them to retain customers and drive additional revenue into their business, rather than answering questions about how many properties they have available to rent at present and in which areas, or how much their fees may be. The real benefit of a chatbot is that they will do exactly as trained. Programmers can input every possible question or scenario, and the chatbot will know what to say. There's no such thing as an 'off day' or forgetting crucial information. Even better, a chatbot can be available 24/7 and can talk to thousands of customers at the same time. This allows all online chats to be recorded and data to be gathered so that organisations can see what questions are being asked and learn more about their customers even though they are interacting with a machine.

Conclusion

Digital transformation is slowly taking effect across the UK property sector, helping to drive open data initiatives and providing a plethora of additional services. Key organisations such as HM Land Registry publicly share access to their data as never before, and are driving digital change with exciting initiatives such as Digital Street.[21] Partners from across the property and digital sectors are working together to explore how emerging technology and innovative uses of their data can help change the property sector for both businesses and consumers.

The potential for PR in proptech is huge: the opportunity is great, and developments are taking place at a rapid rate, but at the same time digital transformation can be met with suspicion, and adoption in many sectors is disappointingly slow. Communication is key to removing this barrier.

Social media is proving an invaluable channel for promoting brands, services, information and collaboration. Equally, content marketing continues to drive traffic and pull in new audiences. Coupled with strong partnerships, the potential for proptech is seemingly unending.

Notes

1 www.theprs.co.uk/news/local-authorities-can-now-levy-and-keep-30000-for-enforcing-property-licensing [Accessed 8 August 2019].
2 There are 4.7 million households in England living in the PRS. The sector has undergone rapid growth over the last ten years and represents 20 per cent of all households in England. https://assets.publishing.service.gov.uk/government/uploads/system/uploads/attachment_data/file/817952/HMOs_and_residential_property_licensing_reforms_guidance.pdf [Accessed 15 September 2019].
3 www.gov.uk/government/publications/houses-in-multiple-occupation-and-residential-property-licensing-reform-guidance-for-local-housing-authorities [Accessed 17 October 2019].
4 www.gov.uk/government/publications/houses-in-multiple-occupation-and-residential-property-licensing-reform-guidance-for-local-housing-authorities [Accessed 17 October 2019].
5 www.legislation.gov.uk/ukpga/2016/22/contents/enacted [Accessed 17 October 2019].
6 www.legislation.gov.uk/ukpga/2018/34/enacted [Accessed 17 October 2019].
7 www.kammadata.com/news/2019/02/the-top-10-trends-for-the-private-rental-sector-in-2019/ [Accessed 17 October 2019].
8 www.kammadata.com/news/2019/10/the-times-focuses-in-on-online-agents-and-unregulated-hmos-and-questions-why-councils-are-not-better-regulating-the-industry/ [Accessed 17 October 2019].
9 www.digitalrisks.co.uk/blog/women-startups [Accessed 15 September 2019].
10 www.kammadata.com/news/page/3/#latest [Accessed 15 September 2019].
11 https://geovation.uk/our-partners/ [Accessed 17 October 2019].
12 https://geovation.uk/accelerator/ [Accessed 17 October 2019].
13 https://hmlandregistry.blog.gov.uk/2019/02/07/digital-transformation-and-the-property-sector/ [Accessed 17 October 2019].
14 Iansiti, Marco and Lakhani, Karim R. *The Truth about Blockchain* (January 2017). Harvard University: Harvard Business Review.
15 https://hmlandregistry.blog.gov.uk/2019/05/16/theres-growing-interest-in-digital-mortgages/ [Accessed 17 October 2019].
16 www.mishcon.com/news/working-with-hmlr-to-trial-uks-first-digitised-end-to-end-residential-property-transaction [Accessed 8 August 2019].
17 www.digitaljournal.com/business/q-a-how-blockchain-is-disrupting-real-estate/article/550195 [Accessed 17 October 2019].
18 https://hmlandregistry.blog.gov.uk/2019/05/24/could-blockchain-be-the-future-of-the-property-market [Accessed 15 September 2019].
19 https://assets.publishing.service.gov.uk/government/uploads/system/uploads/attachment_data/file/662811/HM_Land_Registry_Business_strategy_2017_to_2022.pdf [Accessed 8 August 2019].
20 www.gov.uk/government/publications/introducing-govuk-verify/introducing-govuk-verify [Accessed 8 August 2019].
21 https://hmlandregistry.blog.gov.uk/tag/digital-street/ [Accessed 17 October 2019].

13 Promoting property consultancies

Capitalising on change

Penny Norton

Property consultancies can encapsulate the breadth of the property life-cycle, providing services as diverse as site acquisition and planning through to agency and business rates advice, as well as increasingly operating property assets. Consequently property consultancies communicate with a great diversity of stakeholders.

Invariably, therefore, many property consultancies are large, international and diverse organisations with complex structures. While at first glance the top five[1] may appear to provide a similar service to a similar market, in fact there are significant differences in structure, size and culture: CBRE is a plc with headquarters in the US; JLL is a US headquartered multinational professional services corporation; Savills is a plc with head-quarters in London and offices throughout the world; Knight Frank is a limited liability partnership with headquarters in London but offices globally; and BNP Paribas Real Estate is the property division of BNP Paribas Group, a French-based global banking group.

Despite their differences, external changes both in communications and beyond have presented a new set of common challenges, which this chapter will address.

A strategic approach

How does an organisation with such diversity of functions, staff, locations and customers put in place a PR strategy that provides unity but is relevant to each service line? The approach varies depending upon the organisation's structure and culture, but most property consultancies operate with a single overarching PR strategy, which then determines a range of individual strategies: 'vertical' strategies for individual service lines and 'horizontal' strategies for the communication of a wide range of corporate messages, as depicted in a simplified form in Figure 13.1.

While the overall PR strategy provides the big picture, the communications strategies for each service line contain the detail. The benefit of this approach is that while each service line strategy includes the key corporate objectives and messages contained within the overall strategy, the

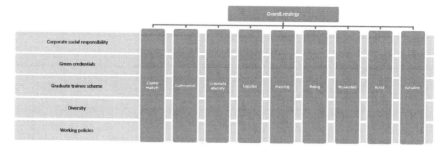

Figure 13.1 Vertical, horizontal and overarching PR strategies.

specific aims and objectives will relate directly to that service line (as shown in Figure 13.2): in this model the overarching strategy is consistent, while the key messages for capital markets vary substantially from those for logistics; likewise, there are few similarities between the target markets for planning and for valuation. This approach also ensures that should corporate changes occur – such as new leadership or organisational expansion – the ongoing PR for each service line is largely unaffected. However, communication though regular meetings and reporting both up and down and across the structure – between the corporate function and the service lines – is essential to avoid conflicting comment: described by one PR director as the 'yin and yang' of property consultancies' various functions,

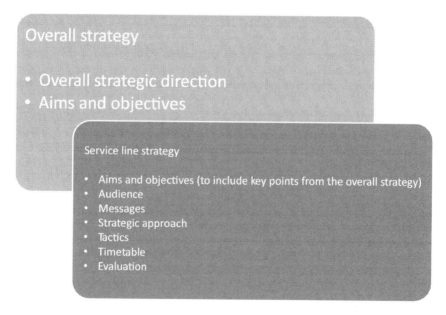

Figure 13.2 Corporate and service line strategies.

an important element of a PR director's role is to avoid situations in which messages communicated on behalf of one service line are detrimental to another.

Strategy is created and implemented primarily by a PR team, the structure of which can vary considerably. Savills' PR team is almost entirely in-house and based in its London headquarters. Individuals, who work under either the B2B or the B2C function, have responsibility both for a service line and for a region but are selected on the basis of their PR skills rather than prior knowledge of a specific sector. Knight Frank, on the other hand, uses an external consultancy, which provides specialist PR support for each of its service lines but uses internal staff for regional PR. LSH, which is particularly strong outside London, uses a combination of in-house and consultancy staff across the country, each serving either the Consultancy Marketing or the Transactional Marketing teams.

PR staff and agencies generally report to a PR or communications director in the headquarters office, though this too can vary, with some regionally based staff and consultancies reporting to a senior director in the relevant regional office. Reflecting the variety of models available, many large consultancies review and change their PR structures on a regular basis, which, despite the obvious complications, can have the benefit of introducing fresh thinking.

The most significant difference between the various models is the extent to which property consultancies use external PR support. Savills opts for an almost entirely in-house team because of the stronger relationships that can be built between PR and the service line heads, greater continuity and control, and the benefits of investing in staff training and development. Those that opt to use external PR consultancies do so because of flexibility, the sector-specific or regional experience and cross-sector experience that consultants can provide, and the opportunity to buy in specific communications functions as required.

There is also variation in how PR is represented at board level, with some PR directors having a role on the board, some reporting to a service line or regional head with responsibility for PR, and others feeding into board decisions though more informal channels, such as regular briefings and occasional attendance at board meetings.

As communications has proliferated substantially, partly as a result of an increasingly broad spectrum of tactics (largely following technological advancements), the PR role has expanded to include new functions, among them search engine optimisation (SEO), social media, video, curation of information using algorithms and the various means of presenting digital information. The newly expanded opportunities for communication, while presenting substantial opportunity, represent a significant challenge to both human and financial resources.

While a clear structure such as that described in Figure 13.1 has many advantages, the fast pace of change and the increasing blurring of distinctions between service lines require agility. Large-scale mixed-use schemes, co-living and build to rent, for example, can potentially involve many of the services that a property consultancy offers.

Property consultancies have overseen a greater proliferation of service lines, with many now taking on the ownership and management of property assets, primarily shared workspaces, but advisory services continue to be the property consultancies' primary offering.

Selling a service

The greatest challenge in promoting the advisory function of property consultancies is, in many cases, the absence of a tangible product. Promoting advice often lacks the clear target of a geographical location or specific business community, and there are few 'milestone' events. When providing property advice, we are rarely involved in the final, celebratory moments of a property's lifecycle; in fact, we rarely come into contact with the end user. Ultimately, we are communicating logic over emotion and expertise over bricks and mortar.

Thought leadership, therefore, is one of the best opportunities to communicate expertise.

Box 13.1 Thought leadership for property consultancies

There is a significant difference between thought leadership and voicing thought. Good thought leadership, rather than adding comment to an already comment-saturated market, is part of a broader campaign that develops thinking or practice, leads by example, and facilitates collaboration and change.

Done well, thought leadership is a great opportunity for a property consultancy to engage with a target audience over a long period of time, to establish niche expertise, to influence the way a market works and to set it apart from the competition.

Selecting a subject/position

- Select a subject that is aligned to your purpose and values, topical, directly related to the service line and of relevance to your target audience – for example, related to social, political, legislative or economic change.
- Try to select a topic (or at least an angle) that has not been addressed by another property consultancy.
- Ideally, identify a subject with broad interest and potentially wider implications.
- Ensure that your subject plays to your strengths.
- Consider precisely how you will add to, and exceed, the existing body of expertise. Check that the corporate PR vision/management is happy with the position advocated.
- Seek potential partners and collaborators.

Developing a thought leadership position

- Understand the change and outcomes that you are looking to achieve.
- Collate new thought on a topic of real interest to your subject – comment opportunity no. 1.
- Turn those insights into practical advice, backed up by example – comment opportunity no. 2.
- Test that practical application in your work – comment opportunity no. 3.
- Implement a collaborative approach that generates broader impact – comment opportunity no. 4.
- Measure its impact.
- Tell the story – comment opportunity no. 5.

Important checks

- The topic and approach should be in line with both the overall and the service line strategy – it must not conflict with key messages and should be relevant to target audiences.
- Your organisation should be abiding by your approach or at least moving towards it.
- Your thoughts should be followed through with actions.
- Thought leadership should not be considered a fixed statement: if the external circumstances are changing, you should constantly review your position in light of this.

The changing role of the property professional

Much has been written about the impact of artificial intelligence (AI) on the property profession.[2] There is no question that data acquisition has become significantly more efficient as a result of technology, and there is concern that this could result in substantial change to some service lines. Now more than ever, property consultancies must become agile to remain viable. As data production demands less of a company's resources than previously, the balance is shifting towards the value that only humans can bring: interpretation, analysis and expert comment on that data. Consequently, the balance between the promotion of comment and statistics is changing.

An increased emphasis on expert comment is not solely an opportunity to create a marketable niche, but a reflection of the proliferation of data to which we are exposed: not all of it reliable, scientific or easy to comprehend. Property professionals are increasingly required to select, curate and make sense of data for the benefit of their clients and to predict its likely impact. This shift in emphasis is also a reflection of the speed of change: whereas previously, ten-year forecasts were the staple PR output of a property consultancy, the pace of political and economic change has rendered such long-term forecasts redundant. In 2005, which property consultancies predicted the worldwide banking crisis and the lengthy period of austerity that followed it? In 2015, which property consultancies predicted the Grenfell Tower tragedy, Brexit, or the inauguration of Donald Trump?

The move online

A centuries-old name is not enough to ensure success in centuries to come; neither is a decades-old approach. As one PR director said, today's environment is one of 'adapt or die': property consultancies have no option but to function online.

It goes without saying that an online service requires online promotion. But the same is true of those services that are still – for the moment at least – primarily offline.

From Valuation to Planning, services are evolving with the proptech revolution: from real-time analytics dashboards and heat maps with live data fields to Building Information Modelling (BIM) and virtual reality.

Possibly the most significant impact of the online revolution is speed. Not only does digital content travel at unprecedented rates, but in a 24/7, information-driven and highly connected environment, expectations for both online and offline resources have increased. The challenge for PR is to ensure the accuracy of the information that is being communicated at speed: never should the race to release information result in a lowering of standards, because in an online world, damage to reputation can also travel at speed.

The nature of the information that is produced is different too: weighty reports are replaced with snippets of information that can be communicated via Twitter; conversely, limitless data *can* be made available – but making it appeal to its intended readership requires considerable PR skill.

The online revolution is essentially re-writing communications theory. The potential for emails and direct messaging on social media to 'go viral' has obfuscated what was a previously a clear option to communicate either one-to-one or one-to-many. Similarly, the decline of traditional media in relation to the potential for a property consultancy to promote news on its website has complicated the 'pull' versus 'push' factors in communicating a message.

Box 13.2 LSH's 574 service: promoting an online product

In January 2019, LSH set up the online property trading platform 574 (www.574.co.uk). 574 has already become a popular auction website for properties across the commercial and residential sectors and has hosted the largest online auction to date.[3]

Promoting the service required a new approach, focused primarily online. Out went the old marketing brochures and flyers for a ballroom-style event at a set date and time. Instead, 574 focuses on digital marketing, SEO and e-marketing, and offers discounts to those signing up to the service online.

The success of 574 has exceeded expectations, attracting over 5,000 online bids in the first six months and appealing to wide-ranging online investors from the institutional, high net worth, professional, private and public sectors.

The content explosion

The amount of data being produced each day is growing exponentially: it is said that over the last two years alone, 90 per cent of the current data in the world has been generated.[4] Partly due to the ease of posting information online, the amount of content produced by property consultancies has increased considerably. A major PR consideration of the content explosion is that the more information that is generated and the greater speed with which it is despatched, the more complex it becomes to protect a reputation. Therefore, this represents both an opportunity and a threat to PR teams: while it is easier to put information into the public domain, information is less likely to be consumed by the target audience – unless it is very well produced, written, presented and promoted.

It falls to PR to curate the abundance of information that a property consultancy produces. Some are moving towards automated curation, which provides the opportunity to present bespoke content via the consultancy's website to users in terms of sector, content and region. Tailoring comment is an effective means of ensuring that clients receive the information that they most need, but we should also take into consideration the 'echo chamber' effect – the fact that people's views can be increasingly limited to the content selected for them by an algorithm – and should aim to find creative means of providing new and interesting information. With cross-selling being one of the most effective means of business development for property consultancies, we should not underestimate the benefits of providing clients with information about service lines that they are not currently using, but in a way that appeals directly to their interests and requirements.

Research: the core PR tool

Many decades-old PR tactics (from hard copy fact files and mail-merged press releases to the all-day lunch) are now obsolete, and a multitude of nascent tools are gaining prominence. But research is the enduring PR tool for the property consultancy.[5] Audiences' insatiable desire for data reflects our increasing opportunity to use data to inform all aspects of life, whether through using comparison websites when making a purchase, league tables to choose a school or Zoopla to inform a conversation about house prices.

Undoubtedly, the intrinsic characteristics of research are well suited to online use and enable it to be adapted to various media channels.

Research enables a property consultancy to demonstrate its understanding of the issues facing its clients and can establish the property consultancy as the expert in the field. In addition to providing directly relevant and unfailingly accurate information, research is most effective if paired with expert comment. Comment should show a deep understanding of the data, demonstrate expertise and be interesting. To issue research to the media and then to find that the journalist has gone to a rival for comment is not a proud moment!

Box 13.3 Research opportunities and examples

Market analysis: CEE Property Investment Market View: Full Year 2019

Market forecasts: The Retail Year: 2019 in Review and Predictions for 2020

Quarterly reports: Global residential investment into London Q4 2019

Predicting market changes: How would an increase in stamp duty impact on the prime residential market?

Industry reactions: Report on Emerging Markets roundtable

League tables: Fit for Work: the office locations with the most bars in a one-mile radius

Maps: Retail parks within ten miles of a major city centre

Impact of new legislation: The impact of permitted development rights on city centre office space

Location analysis: The siting of the next Silicon Valley

Observation of trends: The future of co-working

Social impacts: The impact of major sporting events on local regeneration

Predicting alternative scenarios: The impact of a no-deal Brexit on the industrial and logistics sectors

Future-gazing: Reuse of car parks in the age of the driverless car

---

The most effective research is the 'gift that keeps giving': a set of statistics that tells a national story, provides insight into a variety of sectors, offers both a consumer and a financial angle, can be tailored to specialist and regional press, provides the potential for repeated research at a later date, and, importantly, can be paired with expert comment from a range of voices within the organisation.

The diversification of communications channels has enabled PR to target and repurpose information in a multitude of ways and over a long period of time. And because the services that property consultancies provide are of such broad interest – spanning the consumer, financial and business media – property PR has ample opportunities to repurpose and repackage content, as Box 13.1 shows.

Box 13.4 JLL: research procedures and promotion

JLL's efficient approach to promoting research is based on 'platinum', 'gold', 'silver' and 'bronze' packages – which may be varied depending on the research content – with platinum providing the greatest level of support.

In this case, the topic is identified by the research team and discussed with relevant service lines; then the research is carried out by the research team. A meeting then takes place with the marketing team to discuss the appropriate level of promotion.

Typically, the material is provided to a key publication as an exclusive, and this is followed by an event at which the research findings are unveiled. Bespoke sets of information are then sent to targeted journalists, and research findings are sent to clients and contacts and are posted on the JLL website.

At a later stage, further elements of the research – such as findings relating to specific regions or sectors – are released and the relevant media/contacts targeted.

Where appropriate, the research may be repeated to provide comparative results – for example, at a different point in the property cycle or in a different region – and the process of promotion repeated.

Research can also include organic social media promotion on Twitter, Facebook, Instagram and LinkedIn, and a paid-for digital campaign. Increasingly, video content is used where appropriate.

Box 13.5 JLL: use of maps and infographics in research

In the summer of 2018, JLL seized on the opportunity of the Football World Cup as a topical 'hook' for promoting its expertise.

JLL identified a theme, 'Football's coming home', and carried out a piece of research that identified redeveloped football stadiums across the UK and their impact on the housing crisis.

The research revealed that over 20 years, 15 professional football clubs collectively enabled the delivery of nearly 7,000 new homes, as well as boosting their stadium capacities by an average of 63 per cent. Five London clubs accounted for 3,600 new homes, equivalent to around a fifth of the capital's annual new housing delivery rates.

JLL communicated the findings via a custom-built microsite, which included a map of the UK featuring football grounds. Users were able to zoom in on an area and find out information about the number of homes built, as well as easily accessing comparative information.

The key message was that football clubs can be instrumental in addressing the housing crisis, and specifically that in addition to creating homes, football clubs can provide significant community benefits.

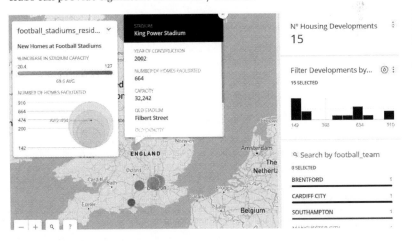

Figure 13.3 JLL's football's coming home website.

Twenty-page reports are obsolete, or at least must be accompanied by a more accessible summary. A director of research for a large property consultancy testifies by The Tube Ride Test: if a research report can be read on a single Tube journey, it is an appropriate length; if not, it is too long. Of course, in a digital context, there are ample opportunities to link to extended reports and additional facts and figures.

The approach is to reverse the old adage: don't let a good story get in the way of the facts. In promoting a piece of research on behalf of a property consultancy, never manipulate facts for a story or be tempted to draw conclusions without the necessary information (be wary of the difference between causation and correlation), never rush to get the information out at the risk of making a mistake, and always provide quality insight in addition to statistics.

PR protocol in a fast-paced media environment

With the opportunity for expert comment comes the risk of erroneous or inappropriate comment and the damage that this can bring to a property consultancy. I am aware of several situations in which has occurred, the most significant one being a planning director slating the introduction of permitted development rights that allowed change of use from offices to homes, much to the detriment of the agency team.

Therefore, protocol (providing it is not over-zealous) and media training is increasingly important, as are briefings by the research or PR teams to media spokespeople that allow contentious subjects to be flagged up, such as those that may cause harm to another service line.

Box 13.6 Protocol: typical content

- Communications principles
- Press releases – corporate/international/UK/local
    template
    checklist
    sign-off procedures – internal/external
    client approval form
    distribution process – online/offline
- Adhering to financial conduct
    quiet period protocol
    publication of financial results
- crisis management
    holding statements
    serious incident reporting
    reputation management
- Social media guidelines
    corporate versus personal accounts
    corporate design guidelines

  advice on effective communication
  online etiquette
  potential pitfalls
- Protocols for arranging media interviews
- Filming guidelines
- Monitoring coverage
- Social media tracker
- Reporting and providing feedback
- Style guide
  written style
  fonts and formatting.

Box 13.7 Media training: a suggested approach for property consultancies

- Identify prior experience of media relations.
- Agree objectives.
- Discuss concerns.
- Discuss how people access news and how this is changing.
- An explanation of how a publication is compiled and the typical working day of a property journalist.
- Discuss 'What is news?' and the importance of information, data, analysis of that data, and surveys.
- Explain the desirable characteristics in a media spokesperson.
- Explain the importance of knowing your audience, journalists' deadlines, and the need for clarity and certainty.
- Advise on having a point of view and sticking to it.
- Explain how exclusives work and how to avoid unwanted media attention.
- Advise on preparing for an interview.
- Explain the importance of corporate 'lines to take'.
- Identify the possible problems that may emerge and how these can be mitigated.
- Practise messaging and interview skills.

Change in the property media

No media has been immune to the online revolution, which has led to the replacement of paper with digital, the rise of blogs, video and podcasts, and the introduction of interactive forums and infographics. In a survey carried out by Cision UK, only 8 per cent of the PR consultants questioned said that they now met journalists face to face, while a third said that social media was the most effective means of contacting the media. Furthermore, 34 per cent of PRs said that journalists were no longer as important to them because of social media,[6] and it has been estimated that journalists have to produce three times as much copy as they did 20 years ago.[7]

The property media, specifically property consultancies' two main printed publications, *Property Week* and *Estates Gazette*, have evolved from weekly magazines to multi-media research, event and content producers, which combine the original publication (somewhat slimmer due to a decline in advertising) with a broad online offering of data, regular bulletins, video and forums. The national media, which has generally followed a similar transformation, tends to allocate less space to B2B property news and consequently has fewer property correspondents. National journalists, who also tend to be younger than previously, cover a broader remit than ever before. A former journalist, now a PR agency director, interviewed for this chapter commented on how the balance of property journalists to PR consultants has shifted from 90 per cent journalists and 10 per cent PRs in the early 1990s to 70 per cent PRs and 30 per cent journalists today, due as much to the decline in the number of journalists employed as to the growth of PR as a profession.

Another PR director commented on how property journalists are so young that few have experienced significant and relevant economic events such as a substantial interest rate rise: she finds that her role increasingly involves explaining facts and figures to journalists and supplying a historical context.

Due to changes in the media, the role of promoting property consultancies increasingly involves writing information that can be copied directly into a news story and in doing so, shaping content to suit a specific publication.

The fast-paced news environment means that carefully timed announcements can be scuppered by the leaking of information online and that the strategy for handling embargoes has changed.

Part of the PR challenge is to communicate this change to those within property consultancies who still believe that coverage in the print edition of the FT is the ultimate PR success. In reality, the FT's global online readership is twice that of its offline readership,[8] and an online message has substantially more potential to be shared.

The aim of simply placing a story in the publication with the highest readership is a blunt instrument in today's tailored media environment. Instead, communication favours more direct channels of communication: a readership of 10,000 directly relevant consumers is more useful than a publication with 200,000 readers of whom only 4 per cent are directly relevant. As a well-respected former journalist commented, senior surveyors no longer aspire to network with a property journalist in preference to a broker with a good Twitter following.

Increasingly, property consultancies are finding that social media, specifically LinkedIn, is the most effective means of publicity. A director of a PR consultancy described a situation whereby he wrote a LinkedIn article on behalf of a surveyor, which was seen and read by thousands of the client's contacts and their networks and ultimately resulted in the property consultancy gaining two new clients to the value of over £1 million. While traditional media can provide the same reach, LinkedIn has the advantage of carrying with it the visible endorsement of key contacts and the opportunity to alert individuals to specific stories.

Social media has also been found to address some of the limitations brought about by GDPR legislation.[9] The Cushman & Wakefield LinkedIn account has over 450,000 followers and is well supported by the company's directors. Frequently, commentary that might previously have been emailed to clients and contacts is posted on the LinkedIn page, promoted by the company's employees to potentially interested parties, and consequently reaches a broad yet relevant audience.

The role of creativity

In the fast-paced media environment there is a tendency for property consultancies to operate on a reactive, rather than proactive, level. Despite best intentions, in a high-pressure environment, strategies and plans can be thrown to the wind. But key messages are not promoted by responding to journalists' requirements: a proactive approach is the only sure way to deliver on the aims and objectives of the strategy.

Property consultancies are generally known for their focus on research and media relations rather than wacky PR stunts. But as the online revolution presents opportunities to do things differently and the content explosion results in greater competition for a share of the voice, PR activity has both increased opportunity for, and a greater need for, creativity and ingenuity. Essentially every traditional PR tactic is adapting, and the increasingly extensive range of communications tools is resulting in tactics being shaped better to their specific use.

Take events, for example. Previously, a set of financial results would have been released at a press conference, which followed a standard format. Today, clients and journalists may be less likely to attend such events due to time pressures and the assumption that the information will be readily available online. Therefore, greater creativity is required to encourage attendance. Property consultancies have achieved this through technological features, such as enabling those present to interact via smartphone and download detailed sets of data that are directly relevant to their specific interest. Property consultancies have also found it necessary to provide a broader range of live, interactive expert comment (via video link if necessary) and to facilitate select seminars on specific subjects within the context of a larger event. Use of social media, such as providing a hashtag in relation to the event, provides broader benefits to clients, who may promote their involvement and their comments to a wider audience, take part in a live forum, or engage in conversation with other attendees at a later stage. The enduring benefits of networking will remain, but this now occurs both offline, online and in a combination of the two.

As property consultancies are increasingly setting up standalone companies such as workspace initiatives, a more entrepreneurial spirit is taking hold. In this field specifically, the target is start-up companies, many of them in the creative industries, and the PR approach must be tailored to the potential client.

Box 13.8 LSH's Airspace Exchange: creativity in a product launch

In the summer of 2019, LSH launched Airspace Exchange (www.airspacexchange.co.uk), a website that provides information about the development value of rooftop space. The initiative followed a change in government policy to allow upwards extensions in certain locations.

LSH's new product, the Airspace Exchange website, enables its user to input property size and location data, and then produces valuation, planning and development advice regarding the possibility of creating additional floors to the building.

The website was promoted with a launch event in a rooftop bar and complemented with a piece of research on space requirements in the UK, a press release, an in-house roadshow, a video showing the untapped potential of rooftop space, and a multi-channel social media campaign linked to the corporate LSH feed.

The launch has exceeded expectations in terms of enquiries generated for the team. Consequently, LSH proposes to introduce a 'chatbot' function to deal with the volume of sales enquiries.

Conclusion

In property consultancy, though media relations is important, it is not the sole purpose of PR. PR, according to the CIPR, is

> about reputation – the result of what you do, what you say and what others say about you. Public Relations is the discipline which looks after reputation, with the aim of earning understanding and support and influencing opinion and behaviour. It is the planned and sustained effort to establish and maintain goodwill and mutual understanding between an organisation and its publics.[10]

The media is simply a means of reaching those publics, and not a public/stakeholder in itself.

As current trends continue, the importance of the traditional media may continue to decline, but the need to communicate with stakeholders will not. Research and expert comment will continue to provide the information that service lines need to communicate with their clients, and PR plays an important role in doing so. And as methods of communication continue to expand, PR must apply increasingly strategic thought, foresight and creativity.

Notes

1 Taken from *Property Week*'s Top European property services firms ranked by UK turnover (July 2018): www.propertyweek.com/agency-and-salary-survey/top-european-property-services-firms-ranked-by-uk-turnover-2018/5097915.article [Accessed 17 October 2019].

2 Thompson, B. and Waller, A. *The Impact of Emerging Technologies on Surveying* (July 2017), RICS. www.isurv.com/info/390/features/11433/technology_future_impact_on_surveying [Accessed 15 September 2019].

3 www.propertyweek.com/private-investor-and-auctions/lshs-574-aims-to-beat-record-lot-price-at-next-online-auction/5102977.article [Accessed 15 September 2019].

4 SINTEF *Big Data, for Better or Worse: 90% of World's Data Generated over Last Two Years* (2013). www.sciencedaily.com/releases/2013/05/130522085217.htm [Accessed 15 September 2019].

5 Researchers will argue, quite rightly, that research is not a PR tool: that most research produced by a property consultancy is commissioned independently of the PR team, and the aspect of promoting research to a wider audience than the client base is simply one aspect of the process of producing and promoting research. Many researchers claim that research is never led, though PR consultants tend to claim the opposite. That said, research is an extremely useful tool when, one way or another, it falls into the hands of the PR consultant.

6 Dayal, Priyanka. *Seven Ways Social Media Changed PR* (May 2015). www.marketingweek.com/seven-ways-social-media-changed-pr/ [Accessed 15 September 2019].

7 www.thetimes.co.uk/tol/life_and_style/education/article7119993.ece, as quoted in Morris and Goldsworthy, *PR Today*.

8 The *Financial Times*' UK monthly global reach for the period 1 January 2017 to 31 December 2017 was 1,129,000 (print), 3,175,015 (web) and 130,094 (app); global figures are 6,002,892 (print), 11,388,235 (web) and 395,354 (app). Source: https://fttoolkit.co.uk/perch/resources/mgrcertificatejanuary-december2017.pdf [Accessed 15 September 2019].

9 The Data Protection Act 2018 prevents property consultancies from sending unsolicited emails without first gaining permission from the recipient. The government's guide to the General Data Protection Regulation is at www.gov.uk/government/publications/guide-to-the-general-data-protection-regulation.

10 www.cipr.co.uk/content/about-us/about-pr [Accessed 8 August 2019].

14 Conclusion

The future of promoting property

Penny Norton and Liz Male

Be silent, or say something better than silence.

(Pythagoras)

The importance of property communications

The significant scope of property sectors featured in this book demonstrates the wide-ranging and fascinating nature of the property industry: from varying property assets such as commercial to luxury residential; to the differing impact of proptech and interior design; or from the knowledge required to promote an emerging community to that required to raise the profile of a property consultancy. Even within these specific areas, the variety of work can be fascinating.

Chapters have focused on the changes facing the property sector, the strategic issues that must be addressed in PR and communications, and the tactics, ranging from community relations to exclusives in glossy magazines, traditional media relations and newsletters, to social media campaigns and digital content curation via algorithm and heat maps.

Commentary and case studies have demonstrated the ability of PR to gain a thorough understanding of many target markets and to engage through wide-ranging and appealing channels; to promote positive messages, expertise, brand awareness and corporate social responsibility, and in doing so, boost asset value, sales and corporate credibility.

This, we hope, leaves no doubt as to what PR can achieve in the property industry while also providing some guidance on the means of doing so.

A sector impacted by complex issues

From our perspective, some of the book's most enlightening and useful content is the detailed investigation of the issues that impact on property PR. PR practitioners questioning the benefits of background research will be left in no doubt as to the need to fully understand not only the product but the issues impacting on it, both internal and external, national and international.

Compared with many other sectors in which PR operates, property is perhaps the most valuable long-term asset. The PR for such a 'big ticket' item should reflect the value of the product. As has been shown, property development (specifically the building of new communities) impacts for decades and even centuries, and goes far beyond simply promoting bricks and mortar. Also very apparent is the emotional attachment to buildings and the places where we choose to live, work and play.

Due to the complexity of the industry and the issues that surround it, the role of educational campaigns and PR activities that help to shape improvements in property and placemaking are of increasing importance – whether in campaigning for increased seating in community spaces or improving the way a property lettings market functions as a result of proptech. Changing legislation and regulation also feature strongly in our experience of the property sector, and this is where PR and public affairs often need to work very closely together.

Unfortunately, a common theme coming out of pre-campaign research is that audiences' understanding of the property sector is often limited, out of date and based on inaccurate assumptions – whether in relation to the impact of change to the built environment, viability in property development or the complexities of the planning process. If the public's understanding of the property industry is limited to the 10 per cent of the iceberg that is above water, our role involves increasing the understanding of the other 90 per cent.

The introduction to this book summarised some key external influences on the property market (Table 1.1), and subsequent chapters have described how that impact can be mitigated or capitalised upon. It is worth noting that none of these factors stands still: politics, stakeholder demands, the campaigning ability of activists, technological development, government regulation, the economic cycle and demographics all continue to change. The astute PR professional will monitor and respond to these changes as quickly as they occur, and should even be able to anticipate them.

Clearly, research is a great basis for any campaign. In reality, though, not all clients have the means to fund it. This is one of the many difficulties facing property PR professionals. Other limitations described in this book include the difficulty in reaching stakeholders (such as future owners of homes yet to be built), ongoing change in property assets, and the use of proptech (not only using technology for PR or promoting proptech, but using proptech to benefit communications itself).

The varying examples described in these chapters demonstrate that while some clients 'get' PR, this is not always the case. Changes to the media environment have shown that the role of PRs as 'gatekeepers' between the client and the journalist is changing. Today, not only do journalists have increased opportunities to access information, but frequently journalists are absent altogether from the process of placing a story into the public domain, as paid, owned and shared media play a larger part in the mix.

PR practitioners are facilitators and must also balance the responsibilities of an ambassador and an impartial advisor. In promoting a company or product, we are clearly its advocate, but we also have a requirement to be objective, often to balance the views of a client so imbued in their product and its benefits that they fail to comprehend external responses to it.

Looking ahead

While the introduction included a section on how property PR has changed in the first two decades of the twenty-first century, it is worth considering how it will continue to change in the future.

First, the product itself is continually evolving. Changing approaches by policy-makers, planners, developers and investors will alter the way in which we describe and promote property. Who would have thought 20 years ago that the word 'luxury' would be used to describe student housing, or that the focus of shopping centres would be as much about experiences as sales? Promoters of student housing increasingly need to focus on the quality of furnishings and smart technology, while those promoting retail will increasingly prioritise interactive art, 'Instagrammable' experiences and store-based digital experiences.

Consumer demands are also changing, not only in relation to the product but also in relation to the expectations of brands to behave ethically and to communicate purpose. The property manager that a developer selects, the practices that the developer carries out internally and even the PR company that it appoints are likely to be increasingly scrutinised. As part of the communications role, we should be prepared to respond to requests for information in relation to diversity, the gender pay gap, wellbeing and sustainability policies.

There is undoubtedly much that the property PR function, specifically the B2B function, can learn from consumer-facing industries such as fast-moving consumer goods (FMCG), retail, leisure and hospitality – and in many areas this is already occurring as a result of integration with these sectors.

And just from a technical point of view, PR strategy must adapt to take into account changes in the media and the emergence of new communications channels, including artificial intelligence (AI), voice technology, virtual reality (VR) and augmented reality (AR), continued changes in traditional media and their business models, accessibility to information held behind paywalls, the increasing micro-targeting of content, changes to the status of social media influencers, new approaches to weeding out fake news and 'deep fake' video and audio content, and many other areas of technological innovation.

The green agenda

> 'Communicators should be helping their organisations or clients to measure what matters, share best practice and case studies, helping to contribute to an industry that can make significant strides in reducing a country's overall environmental impacts.'
>
> Dan Gerrella – chair of CAPSIG (the CIPR's construction and property group)

Calls for greater environmental sustainability will continue to shape the function of property PR as the issues surrounding climate change become more urgent and greater regulation is brought about as a result. It will be increasingly necessary for PRs to understand the issues, the products, services, legislation and standards relating to sustainability. This includes using the right terminology, ensuring that information is accurate and does not mislead audiences – there can be no more greenwash. It also means being a strong advocate at board level to encourage property organisations to step up to probably the biggest and most urgent issue of our time.

The online revolution

> 'The use of technology has enabled us to understand our publics better, to understand how they think and feel about development in their area and to monitor conversations online. This in turn has helped inform better and more robust PR programmes. And while further automation and artificial intelligence will continue to aid the speed and effectiveness of the analysis and interpretation of data, the application of the information including unique idea generation for communication campaigns affecting social issues (such as housing) continues to require a human touch, for now at least. I'd like to look back on this statement in 5 or 10 years and see if it is still true then!'
>
> Emma Drake – founder, Henbe

The significant change to communications brought about by online and mobile technology cannot be overestimated. PR strategy, tactics, resources and evaluation will all continue to change in the face of the digital revolution. The traditional models of communication, such as the four models of communication (press agentry, public information, two-way asymmetrical and two-way) espoused by Grunig and Hunt,[1] will require updating to fit twenty-first century communications.

As automation takes on an increasing role, many aspects of property PR, including research, targeting, communications tactics, dissemination of information, monitoring, analysis and evaluation, will be carried out with input from AI. But what about the role of the property PR practitioners themselves?

In 2018, the CIPR launched a crowdsourced literature review to explore the impact of AI[2] based on the assumption that '[s]oftware will increasingly be used to create content; content marketing will be driven by algorithms; bots will manage public enquiries; and decisions of channels and tactics will increasingly be automated, driven in real time by public responses and behaviours.'

The report concluded that copywriting, strategic planning and social media are all likely to be impacted by AI. Although positive about the usefulness of AI, the report states that AI can be detrimental:

> We need humans to think creatively and abstractly about problems to devise new and innovative strategies, test out different approaches and look to the future. Parts of what we do – or in some cases entire tasks – are or will be automated and will benefit from AI. Regardless of the tasks and skills that can be automated or benefit from AI, human intervention in editing, sensitivity, emotional intelligence, applying good judgement and ethics will always be needed … Education, experiential learning and continuous development of these very human traits that are valued in our profession.

There remains a human element to writing news stories and other content, which is currently beyond the capabilities of even the most advanced AI systems. But apparently this will not necessarily remain the case. The report states that currently, as much as 12 per cent of PR has been replaced by AI, primarily in the context of evaluation, data processing, programming and curation. It states that the figure is likely to rise to 36 per cent by 2023, taking in various functions, from stakeholder analysis to reputation monitoring.

Clearly, PR professionals must keep on their toes to fully understand both the benefits and pitfalls of AI, to accept the inevitable changes and to use them to the benefit of our work.

Changing media relations

> 'Social media and the digital world have changed everything in terms of spreading messages, profiles, stories and crises. It used to take three weeks for a story to land; now it's minutes, and you can see it tripping across the world in hours. You have to be on the ball and ready to react any time of the day … and night.'
>
> Jamie Jago – head of Residential PR, Savills

The introduction and subsequent chapters of this book have described the changing world of the property media. As with many forms of communication, online capabilities create opportunities for media relations to operate as a two-way, symmetrical tactic as opposed to an information-only tactic – an opportunity that good PRs have benefited from for years. Podcasts, webinars, blogs, vlogs and discussions on Twitter all enable extended dialogue, which was previously beyond the capabilities of traditional media relations.

Although always important since the syndicated tapes of the 1980s, video is particularly effective online. It has been predicted that video will account for over 80 per cent of all internet traffic by 2021,[3] partly as a result of live video streaming apps.

Private messaging apps now outperform the major social media networks in terms of active users: an estimated one billion people use WhatsApp.[4] Business are increasingly delivering bespoke content in this way, with carefully targeted newsletters, videos, information and offers.

Evaluation

'There is increasing measurement of success. As PR becomes more digital, agencies are able to measure every step of the communications process, from output to outcomes. Using real-time data, insights, we're better able to understand how our messages are received. Using all the tools at our disposal we have access to comprehensive online, print, broadcast and social media monitoring and analysis.'

Dan Innes – managing director, Innesco

It would be impossible to determine the success of any of the innovations described earlier without evaluation – although some clients and even some PR professionals appear to disagree. Evaluation is a key component of strategic PR; it enables us to develop our skills and demonstrates the effectiveness of our work.

As PR becomes increasingly digital, we are increasingly able to measure every step of the communications process, from output to outcomes, using tools such as Google Analytics and social media metrics to provide real-time data and sophisticated insights.

Furthermore, substantial advances in evaluation approaches have been made as a result of the Barcelona principles,[5] AMEC'S Integrated Evaluation Framework[6] and the substantial body of work carried out by the CIPR.[7]

Changing roles

'The role of public relations is changing across the profession and this is no different in the property sector. We still have a job to do as practitioners, to educate and inform clients and colleagues on the role of best practice PR, not just in media relations but as a management discipline that can support the trust, relationships and reputation of companies or projects with their publics.'

Emma Drake – founder, Henbe

Today's PR consultants are brand ambassadors, social media experts, data analysts, content marketers, crisis managers, boardroom advisors and trend spotters in addition to the tasks associated with their traditional role, and there is undoubtedly greater integration between PR, digital and marketing functions. Taking into account the wide range of responsibilities, communications tools and tactics that are currently linked to property PR, or are likely to be used in the future, there is likely to be a further blurring of the roles – and possibly a suggestion that 'PR' as a definition is out of date.

As this book has shown, there are still pockets of property PR practice where the more traditional, media-relations-centric PR of the twentieth century is alive and kicking. Maybe this will always remain, reflecting the fundamental importance of human relationships in a sector that is very sociable and people focused.

Traditional PR skills are enduring – understanding the audience and the business context and using research to inform a strategy, creating a positive message and making it stand out; striking the delicate balance between what the client/property developer wants to promote and what the audience wants to hear; using creativity to select and implement appropriate (and genuinely engaging) tactics; and measuring the outcomes of all this for business performance, awareness and corporate reputation. All these are as relevant today as they ever have been.

Set alongside this is also an increasing understanding and use of PR as a strategic management discipline operating at board level, an increasingly sophisticated understanding of digital communications, and a much-improved approach to evaluating impact and demonstrating return on investment (ROI).

If, as we believe, PR is the cornerstone in the relationship between an organisation and its publics, it undoubtedly has an important role in future communications for all parts of the property sector.

> 'Public Relations is about reputation – the result of what you do, what you say and what others say about you. Public Relations is the discipline which looks after reputation, with the aim of earning understanding and support and influencing opinion and behaviour. It is the planned and sustained effort to establish and maintain goodwill and mutual understanding between an organisation and its publics.
>
> 'Every organisation, no matter how large or small, ultimately depends on its reputation for survival and success. Customers, suppliers, employees, investors, journalists and regulators can have a powerful impact. They all have an opinion about the organisations they come into contact with – whether good or bad, right or wrong. These perceptions will drive their decisions about whether they want to work with, shop with and support these organisations.

'In today's competitive market, reputation can be a company's biggest asset – the thing that makes you stand out from the crowd and gives you a competitive edge. Effective PR can help manage reputation by communicating and building good relationships with all organisation stakeholders.'

The CIPR[8]

Notes

1 Grunig. *Excellence in PR*.
2 The project, #AIinPR, was led by Anne Gregory, Professor of Corporate Communication at the University of Huddersfield and a former president of the CIPR. The report, and other relevant information, is available at https://newsroom.cipr. co.uk/aiinpr-panel-publishes-introduction-to-ai-in-pr/ [Accessed 9 January 2020].
3 https://oxygenagency.co.uk/marketing-trends-2019/ [Accessed 8 September 2019].
4 https://heimdalsecurity.com/blog/the-best-encrypted-messaging-apps/ [Accessed 8 September 2019].
5 For a more detailed explanation, see Chapter 3 (endnote 9) or https://amecorg. com/barcelona-principles-2-0/ [Accessed 9 January 2020].
6 https://amecorg.com/amecframework/ [Accessed 11 September 2019].
7 www.cipr.co.uk/content/policy-resources/toolkits-and-best-practice-guides [Accessed 11 September 2019].
8 Ibid [Accessed 11 September 2019].

Glossary

Above the line Promotional methods that cannot be directly controlled by the company selling the goods or service, such as television or press advertising (see Below the line).

Accessibility Freedom for people to take part, including elderly and disabled people, those with young children and those who may encounter discrimination.

Accessible housing Units designed to allow easier access for physically disabled or visually impaired people.

Ad-blocking A piece of software to prevent advertisements from appearing on a web page.

Affordable housing Social rented, affordable rented and intermediate housing, provided to eligible households whose needs are not met by the market. Eligibility is based on local incomes and local house prices. (UK)

Algorithm A method used by search engine providers to find and rank online content.

Allocated site A site with potential for development and allocated as such in a local authority's Local Plan.

ALMO Arms-length management organisation: a not-for-profit company that provides housing services on behalf of a local authority. (UK)

Amenity A positive element that contributes to the overall character or enjoyment of an area. For example, open land, trees, historic buildings, or less tangible factors such as tranquillity. Residential amenity considerations may include privacy (overlooking), overbearing impact, overshadowing or loss of daylight/sunlight.

Analytics The data collected about visitors to a website; used to understand user behaviour. Also used as shorthand for Google Analytics, a specific service provided by Google.

App (application) A type of software program that can be downloaded onto a computer, tablet or smartphone.

Appeals The process whereby a planning applicant can challenge a decision, usually refusal of planning consent. Appeals can also be made against the failure of a planning authority to issue a decision within a given time and against conditions attached to a planning permission. In England and Wales, appeals are processed by the Planning Inspectorate.

Area Action Plan A type of Development Plan Document focused upon a specific location or an area subject to conservation or significant change (for example a major regeneration scheme) (UK).

Area of Outstanding Natural Beauty (AONB) An area with statutory national landscape designation, the primary purpose of which is to conserve and enhance natural beauty (UK).

Artificial intelligence The use of computer systems to perform tasks normally requiring human intelligence, such as visual perception, speech recognition, decision-making and translation between languages.

Asset management A systematic approach to the governance and realisation of value from a property asset.

Audience/target audience A specified group defined for marketing purposes.

Authentication Online, the process of verifying a user's identity prior to a transaction, such as a domain name transfer.

Automated curation Use of IT systems to discover, gather and present digital content on a specific subject.

Average session time The total duration of all sessions (in seconds) spent on a website, divided by the total number of sessions.

B2B Business to business PR.

B2C Business to consumer PR.

Barcelona Principles A series of statements to guide best practice in PR measurement that were endorsed after a vote of global delegates at the AMEC European Measurement Summit in 2010.

Bedroom tax An informal name for a measure introduced in the UK Welfare Reform Act 2012, by which the amount of housing benefit paid to a claimant is reduced if the property they are renting is judged to have more bedrooms than necessary.

Below the line Promotional tactics that can be controlled by the company selling the goods or service, such as in-store offers and direct selling (see Above the line).

BID See Business Improvement District.

BIM See Building Information Modelling.

Blockchain A system in which a record of transactions made in bitcoin or another cryptocurrency is maintained across several computers that are linked in a peer-to-peer network.

Blog An abbreviation of weB LOG: a journal that is available online and is updated by the owner regularly.

BME Black and Minority Ethnic: terminology used to describe people of non-white descent. Also often referred to as BAME.

Bot A software application that runs automated tasks over the internet.

Bounce rate The percentage of visitors to a website who navigate away from the site after viewing only one page.

Brand   The tangible and intangible attributes of a product or organisation that create an image in the public mind.

Brand associations   The knowledge and feelings that consumers associate with a brand name.

Brand collaboration   A strategic tool used to gain higher profits through an alliance with another powerful brand name.

Brand essence   The intangible characteristics that define a brand.

Brand tone of voice   A description of how the brand speaks to consumers.

BREEAM   Building Research Establishment Environmental Assessment Method – a recognised environmental assessment method and rating system for buildings in the UK.

Broadband   A high-speed internet connection.

Brownfield land   Land that is or was occupied by a permanent structure.

Browser   Computer software that can be used to search for and view information on the internet.

Build to rent (BTR)   A term used to describe an emerging, fast-growing market for residential property, designed for rent. Developments are typically owned by companies (such as property companies or pension or insurance investment companies) and let directly or through an agent.

Building Information Modelling (BIM)   An intelligent 3D model-based process used to plan, design, construct and manage buildings and infrastructure.

Buildings standards   A set of standards established and enforced by local government for the structural safety and broader performance of buildings (UK). Also Building Regulations.

Bulletin board (online)   An online noticeboard.

Business Improvement District (BID)   Designated town centre (and sometimes other areas) scheme whereby businesses agree to pay additional rates to fund improvements to the general retail environment (UK).

Buy-to-let   A product bought by a landlord with the intention of renting it to tenants.

Call-in   In the UK, the Secretary of State for Housing, Communities and Local Government can order that a planning application or Local Plan is taken out of the hands of a local authority. The application will then be subject to a public inquiry presided over by a Planning Inspector, who will make a recommendation to the Secretary of State.

CGI   Computer-generated imagery – special visual effects created using computer software.

Change of use   In the UK, planning permission is required for the 'material change of use' of a property except in specific circumstances – most recently following Permitted Development Rights legislation.

Chatbot   Short for chat robot, a computer program that simulates human conversation, or chat, through artificial intelligence. See Bot.

CIL   See Community Infrastructure Levy.

CIPR   Chartered Institute of Public Relations – the professional body for the UK public relations industry, providing training and events, news and research.

Circulation   The total number of copies of newspapers, magazines or other print publications distributed by a specific print publication.

CMP   See Construction Management Plan.

Code for Sustainable Homes   A now abandoned UK national standard for sustainable design and construction of new homes.

Co-living   An umbrella term for co-housing where two or more people who are not related live together.

Commercial property   Properties that include retail, office buildings, hotels and service establishments, although often used specifically in relation to office property.

Community benefits   Aspects of a proposed development that bring about social, economic or environmental benefits. Community benefits may be put in place to mitigate the impact of development.

Community development fund   A sum of money set aside by a developer for the benefit of the community.

Community engagement   Activities undertaken to establish effective relationships with individuals or groups within a defined community.

Community facility   An asset provided for the benefit of the community, such as a community centre, church or library.

Community Infrastructure Levy   A planning charge, introduced by the Planning Act 2008, as a tool for local authorities in England and Wales to deliver infrastructure to support development.

Community involvement   Effective interactions between applicants, local authorities, decision-makers, and individual and representative stakeholders to identify issues and exchange views on a continuous basis.

Community liaison officer   A person who liaises with the community on behalf of either a local authority or a development team to enable the two organisations to communicate and work together.

Community Relations   Social outreach programmes designed to build relations and foster understanding of the role of the business with neighbours in the local community.

Compulsory Purchase Order (CPO)   An order issued by government or a local authority to acquire land or buildings in the wider public interest (UK).

Conservation   The process of maintaining and managing change to a heritage asset in a way that sustains and, where appropriate, enhances its significance.

Conservation area   An area of special architectural or historic interest, the character, appearance or setting of which it is desirable to preserve or enhance.

Conservation society An organisation established with the prime objective of protecting and preserving the environment. Typically, conservation societies exist to protect original architecture or natural resources.

Construction Impacts Group A group set up, often by a developer, to mitigate the negative impact of development on a community.

Construction Management Plan A set of conditions put in place by a local authority to ensure that developers minimise the negative impact of construction.

Consultation The process of sharing information and promoting dialogue between local planning authorities, applicants, individuals or civic groups, with the objective of gathering views and opinions on planning policies or development proposals.

Consultation fatigue The reluctance to take part in consultation, usually as a result of excessive past consultation or lack of demonstrable results from previous consultation.

Conveyancing The legal process of buying and selling land.

Core Strategy A Local Plan document that sets out the long-term vision, strategic objectives and strategic planning policies for a local authority area.

Corporate Social Responsibility (CSR) The recognition that a company or organisation should take into account the effect of its social, ethical and environmental activities on its staff and the community around it.

CPO See Compulsory Purchase Order.

Crisis Communications A damage limitation communications process used by organisations when experiencing a crisis.

Cross-selling To sell a different product or service to an existing customer.

CSR See Corporate Social Responsibility.

Demographics The social and economic characteristics of a group of households or individuals.

Developer trust A fund put in place by a developer to provide an ongoing financial resource to the community, usually for a specific purpose.

Digital divide The gap between those who have access to technology and those who do not, due typically to availability of technology and network coverage but also money, location or literacy.

Direct marketing Advertising and printed promotional material such as brochures, flyers and mailshots sent directly to customers.

Discussion board (online) An online 'bulletin board' where individuals can post messages and respond to others' messages.

Discussion forum (online) See Discussion board (online).

Discussion group (online) See Discussion board (online).

District plan A document outlining a local authority's plans for the management of land.

Dwell time The amount of time that a user spends on a website.

Echo chamber A metaphorical description of a situation in which beliefs are amplified or reinforced by communication and repetition inside a closed system.

Embargo The sharing of unannounced, relevant information between a communications professional and the media, with agreement that the information cannot be published before an agreed-upon time and date.

Energy Act (The) UK legislation focused on improving the energy efficiency of buildings.

Energy Performance Certificate (EPC) Energy efficiency ratings for residential properties (UK).

Evaluation The continuous process of measuring the impact of a PR campaign.

Exclusive A news story offered by a PR practitioner to a single newspaper title, radio, website or TV station.

Fake news A recent term used to describe false stories that appear to be news, spread on the internet or using other media.

FCA The Financial Conduct Authority – a financial regulatory body in the UK.

Fintech Computer programs and other technology used to support or enable banking and financial services.

Footfall The number of people visiting a shop or business in a particular period of time.

Freehold Outright ownership of a property and the land on which it stands.

Generation X The generation born during the 1960s and 1970s.

Generation Y The generation born in the 1980s.

Generation Z The generation born in the 1990s.

Gentrification The process by which wealthier people move into, renovate and restore deteriorated areas.

Geo-marketing The use of geographic location information in marketing promotions.

Geo-targeting The process of targeting a marketing or advertising campaign at a limited set of consumers based on their physical location.

Google Analytics A web analytics service that tracks and reports website traffic.

Googlemaps A web mapping service that offers satellite imagery, street maps, 360° panoramic views of streets, real-time traffic conditions and route planning for travelling by foot, car, bicycle or public transportation.

Grassroots engagement Bottom-up, rather than top-down, decision-making, sometimes considered more natural and spontaneous than more traditional power structures.

Greenbelt A designation for land around certain cities and large built-up areas, which aims to keep the land permanently open or undeveloped.

Greenfield Previously undeveloped land.

Ground rent   The annual charge levied by a freeholder on a leaseholder of a property.

Hard-to-reach groups   Those groups of society which it is particularly difficult to communicate with through the usual means.

Heat maps   A graphical representation of data where the individual values contained in a matrix are represented as colours.

Help to Buy   A UK government scheme designed to help anyone struggling to save a deposit for their first home or move up the property ladder.

High net worth   An individual who has self-certified that during the financial year immediately preceding the date in question, they have an annual income to the value of £100,000 or more.

High-end   A business that makes or sells expensive products.

HLA   See Housing Land Availability.

HMTL   Hypertext Markup Language: a standardised language of computer code, imbedded in 'source' documents behind all web documents, containing textual content, images, links to other documents and formatting instructions for display on the screen.

Housing association (HA)   A common term for independent, not-for-profit organisations that work with local authorities to offer homes to specific demographics at a reduced cost.

Housing Land Availability   The total amount of land reserved for residential use awaiting development (UK).

Housing need   A level of socially desirable housing, the demand for which is not reflected in the open market.

Hyperlink, hypertext   Text on a web page that links to another document or web page.

Impressions   In online analytics, a term to describe the number of times that content is displayed (for example on social media or a website).

Influencers   Bloggers, journalists and companies who are thought leaders in their industry and recognised by customers as people to trust.

Infographic   An image that breaks down the facts or messages around a key subject into simple graphics.

In-house   Staff within a company or organisation (in PR, the term is used to differentiate from consultancies).

Inspector's report   A document produced by an independent inspector from the Planning Inspectorate. It assesses the soundness and robustness of planning documents (UK).

Integrated campaign   Use of multiple marketing communications channels such as online, print, TV and radio, B2B, direct marketing, video and advertising.

Integrated communications   The linking together of all forms of communications and messages.

Issue   A matter of concern with potential to become a crisis.

Issues log   A simple list or spreadsheet that helps managers track the issues that arise in a project and prioritise a response to them.

Issues management   Ongoing activity that includes studying public policy matters and other societal issues of concern to an organisation.

Key performance indicator   A set of values against which to measure success; must be defined to reflect objectives and strategy and be sufficiently robust for the measurement to be repeatable. KPIs can be presented as a number, ratio or percentage.

KPI   See Key performance indicator.

Land Registry   The government department that maintains the national property register in England and Wales.

Lease   The legal document governing the occupation by the tenant of premises for a specific length of time.

Leasehold   The use and occupation of a property by way of a lease agreement for a certain period of time. A lease is frequently applicable to flats but can also apply to houses.

Letting agent   A person (usually a surveyor) acting for the party seeking to let a property in the open market.

Listed building   A building of special architectural or historic interest. A listed building may carry certain obligations and restrictions governing its use, repair and maintenance.

Live data fields   Online information updated instantaneously.

Lobbying   The process whereby individuals, civic groups or commercial organisations seek to influence planning decision-makers by employing a variety of tactics.

Local authority (LA)   Local government in England consists of five different types of local authorities: single-tier (metropolitan authorities, London boroughs, unitary or shire authorities) and two-tier authorities (county council and district council). The nearly 400 local authorities are responsible for a range of services for people and businesses in a defined area and are made up of permanent council staff, council officers, and elected councillors. See also Local planning authority.

Local Plan   The main planning policy document for a local authority area. A Local Plan's 'development plan' status means that it is the primary consideration in deciding planning applications. (UK)

Local planning authority   The public authority whose duty it is to carry out specific planning functions for a particular area. In the UK this includes district councils, London borough councils, unitary authorities, county councils, the Broads Authority, the National Park Authority and the Greater London Authority.

Localism   The 2011 UK Localism Act devolved greater powers to local government and neighbourhoods and gave local communities additional rights over planning decisions.

Low carbon homes   Buildings specifically engineered to achieve greenhouse gas reduction.

Marketing channels   The people, organisations and activities necessary to transfer the ownership of goods from the point of production to the point of consumption.

Masterplan   A document outlining the overall approach to the layout of a development.

Media relations   Communications with the news media on behalf of an organisation.

Messages   Agreed words or statements that an organisation intends to communicate to its audiences.

Messaging strategy   A set of foundational points that are aligned with a company's goals and overall brand messaging.

Metrics   The use of data to gauge the impact of activity, for example on a website or on social media, used to gather information about how a brand, product or company-related topic is perceived.

MHCLG   See Ministry of Housing, Communities and Local Government.

Microsite   A small auxiliary website designed to function as a supplement to a primary website.

Micro-targeting   Direct marketing data-mining techniques that involve predictive market segmentation.

Ministry of Housing, Communities and Local Government (MHCLG)   The UK government department with responsibilities for housing, planning and development.

Mixed use   Development projects that comprise a mixture of land uses.

Mixed-use development   Developments constituting more than one use type.

Mobile-optimised sites   Websites adjusted to ensure that visitors accessing the site from mobile devices have an experience customised to their device.

Monitoring   Regular measurement of progress towards targets, aims and objectives. Also involves scrutiny, evaluation and, where necessary, changes in policies, plans and strategies.

New Towns   New Towns in the UK were planned under the powers of the New Towns Act 1946 and later Acts to relocate populations in poor or bombed-out housing following World War II. Designated New Towns were removed from local authority control and placed under the supervision of a development corporation. The corporations were later disbanded and their assets split between local authorities and, in England, the Commission for New Towns (later English Partnerships).

News angle/news hook   Information that is new, important, different or unusual about a specific event, situation, or person and which makes it newsworthy.

NHBC   National House Building Council, which provides homeowners with a ten-year warranty against major structural defects for new properties (UK).

NIMBY An abbreviation of Not In My Back Yard, used in relation to those who oppose development in the vicinity of their homes for purely selfish reasons.

Notification An online alert in relation to online activities.

Objection A written representation made to a local planning authority by an individual, civic group or statutory consultee in response to Local Plan proposals or a planning application.

Ombudsmen Independent professional bodies that investigate complaints on behalf of customers.

Online consultation Consultation that takes place via website, email, social media or other online means.

Online forum See Discussion forum.

Op-ed A newspaper article written by an expert that is positioned on the page opposite the editorial page.

Open data Data freely available for everyone to use and republish as they wish, without restrictions from copyright, patents or other mechanisms of control. data.gov.uk is a UK government project to make available non-personal UK government data as open data.

Open house (or open viewing) A process, normally managed by an estate agent, whereby several potential purchasers are given a set time during which they may view a property for sale instead of separate, private viewings.

Open/public meeting A meeting (for example to launch a consultation or present and discuss a development proposal) that is open to all.

Organic search The method of finding a website by entering search items into a search engine.

Outline application An application for planning permission that does not include full details of the proposal. Outline consent approves the principle of development, and detailed consent is provided at a later stage.

Page views The number of times a web page was viewed.

Paid media Traditional advertising, also advertising online.

Pepper-potting A form of mixed-tenure development in which poorer and more affluent residents live in a mixed community through the 'sprinkling' of social housing among privately owned housing.

Perception audit/survey A strategic marketing research technique to provide a baseline indication of existing attitudes against which to measure future marketing efforts.

Permitted development Permission to carry out (usually limited) forms of development without the need for a planning application. In the UK these provisions are granted under the Town and Country Planning (General Permitted Development) (England) Order 2015. Local planning authorities have the power to remove permitted development rights through planning conditions or Article 4 Directions.

PEST analysis   A framework of macro-environmental factors used in strategic management, which gives an overview based on political, economic, social and technological factors.

Photo call   Advance notice to the media of a formally organised opportunity at a set time and date to take a press photograph of a particular person or event.

Placemaking   A multi-faceted approach to the planning, design and management of public spaces, which capitalises on a local community's assets, inspiration and potential, with the intention of creating public spaces that promote people's health, happiness and wellbeing.

Planning committee   The planning decision-making body of a local authority. In the UK the planning committee is made up of elected members. Its main role is to make decisions on planning applications.

Podcast   A digital audio file made available on the internet for downloading to a computer or portable media player.

Pop-up events   Temporary events, often held in unconventional locations.

Portal   A term, generally synonymous with gateway, for a website intended to be a starting site for users when they connect to the internet.

Pre-application discussions   Meetings between a prospective applicant and a local authority prior to making a planning application, generally confidential in nature.

Pre-election purdah   The period of time between the announcement of an election and the results. Frequently requires a suspension of any activity that may influence the result.

Preferendum   A consultation technique, similar to a referendum, that uses a selection of options in place of yes/no questions.

Press release   A written communication sent to all news media, usually put out by a representative of a company, organisation or individual.

Press tour   Coordinated visits by PR professionals to secure multiple media opportunities.

Previously developed land   Land that is or was occupied by a permanent structure. See also Brownfield land.

Print circulation   The total number of copies of a publication available to subscribers as well as via newsstands, vending machines and other delivery systems.

Print production   The process of producing printed material such as brochures, posters and leaflets.

Prior approval   A procedure whereby permission is deemed granted if the local planning authority does not respond to the developer's application within a certain time.

Private Finance Initiative (PFI)   The Private Finance Initiative in the UK was developed as a means of creating public–private partnerships (PPPs) whereby private firms are contracted to complete and manage public projects.

Project liaison group A group of stakeholders, usually representative of the wider community, with whom the development team discusses a development proposal on an ongoing basis.

Property Ombudsman (The) A free and independent service for resolving disputes between those sales and letting agents who are members of The Property Ombudsman and buyers/sellers of residential property in the UK.

Proptech Property technology: the use of information technology to help individuals and companies research, buy, sell and manage real estate.

Public affairs The process of communicating an organisation's point of view on issues or causes to political audiences, including MPs and lobbying groups.

Public realm The external spaces in towns and cities that are accessible to all.

Public sector All public services in the UK, including emergency services, healthcare, education, social care, housing and refuse collection.

Publics Target audiences of a company, organisation or individual.

Qualitative research A scientific method of observation to gather non-numerical data.

Quantitative research A scientific method of analysing data via statistical, mathematical or computational techniques.

Reach A data metric that determines the number of people (or percentage of an audience) that have been exposed to content.

Real-time analytics dashboards Visualisations that are automatically updated with current data and can be tailored to provide the most relevant operational reporting data.

Real-time data Information that is delivered immediately.

Regeneration The use of public money to reverse decline by improving the physical structure, local economy and social/community infrastructure.

Registered land Land that, in the UK, has its ownership details recorded at the Land Registry.

Registered social landlord (RSL) Social landlords that in England were formerly registered with the Housing Corporation, or in Wales with the Welsh Government.

Reputation management The PR practice of monitoring, correcting and enhancing the perception of a brand, individual, organisation or business in the public's opinion.

Return on investment (ROI) A performance measure used to evaluate the efficiency of an investment.

Retweet Used in social media: a Twitter user endorses another Twitter user's tweet by forwarding it to their network.

Risk management Preventive PR whereby an organisation focuses on identifying areas of potential danger and mitigating against a crisis.

ROI See Return on investment.

RSL   See Registered social landlord.

Scatter-gun   A way of doing something in a way that is not well organised. The scatter-gun approach in PR is not targeted at particular individuals.

Search engine optimisation (SEO)   Producing greater visibility for a website by planning and adjusting the content, keywords and phrases of a web page in order to improve its position in search results (search engine ranking).

Section 106 obligations   Requirements of developers as part of planning permissions. These are agreed in the planning application process to provide contributions (usually financial) to develop facilities/amenities for the local community (UK).

Sell-in   The process of communicating a news story or idea to a journalist.

Sentiment   The positive or negative feelings and/or perceptions a public has about a given subject.

SEO   See Search engine optimisation.

Share of voice   An advertising revenue model that focuses on weight or percentage as compared with that of other advertisers.

Shared ownership   A means by which a resident owns a share of a property and pays rent on the remaining portion.

Shoppertainment   A buzzword used by retailers to describe an overall in-store experience designed to bring customers back into brick-and-mortar stores.

Silent majority   A significant proportion of a population who choose not to express their views, often because of apathy or because they do not believe that their views matter.

Site acquisition   Obtaining land for development purposes.

Site allocation   Designation of land in local development document for a specific land use.

Site appraisal tools   A means of appraising potential sites for development.

Site of Special Scientific Interest   Land designated for special protection by Natural England under the Wildlife and Countryside Act 1981 (UK).

Site Specific Allocations   The allocation of sites for specific or mixed uses. Policies will identify any specific requirements for the site (UK).

Situational analysis   A collection of methods that can be used to analyse the internal and external environment in relation to a specific proposal. See also PEST analysis and SWOT analysis.

Smart energy systems   Cost-effective, sustainable and secure energy systems in which renewable energy production, infrastructures and consumption are integrated and coordinated through energy services, active users and enabling technologies.

SMS   Short Message (or Messaging) Service: a system that enables mobile phone users to send and receive text messages.

Social aggregation sites   Websites that collect content from multiple sources and re-present it in one location.

Social analytics Search, indexing, semantic analysis and business intelligence technologies used in identifying, tracking, listening to and participating in the distributed conversations about a brand, product or issue, with emphasis on quantifying the trend in each conversation's sentiment and influence.

Social housing Housing provided for those on low incomes by local authorities, government agencies or non-profit organisations.

Social landlords Those who own and manage social housing (usually councils or housing associations). Surpluses are re-invested in managing and maintaining existing homes, providing associated services and, in some cases, building new homes.

Social media Websites and applications that enable users to create and share content or to participate in social networking.

Social responsibility Providing corporate resources to demonstrate an organisation's commitment to ethically responsible behaviour.

Soundbite A short clip of speech, often used to promote or exemplify the full-length piece.

Special interest group A community with a shared interest, where members cooperate to affect or to produce solutions in relation to their area of specific interest and may communicate, meet and organise events.

Spin A derogatory term for the act of highlighting the positive aspects of a bad situation, statement or action, usually to the news media.

Spokesperson A person who is selected and trained to speak on behalf of a company, organisation or brand.

SSSI See Site of Special Scientific Interest.

Stakeholder A person or group with an economic, professional or community interest in an organisation's activities.

Stakeholder/publics analysis The process of analysing stakeholder groups and their (likely) views.

Stakeholder engagement The process by which an organisation involves people who may be affected by the decisions it makes or can influence the implementation of its decisions.

Stakeholder mapping The process of identifying stakeholder groups geographically.

Stamp Duty A tax paid to the UK government by a homeowner upon completion of a sale.

SWOT analysis An acronym for strengths, weaknesses, opportunities and threats; a structured planning method that evaluates those four elements of a project or business venture.

Syndicated A news story placed on several websites or in several outlets.

Target audience The groups to be targeted as recipients of a message.

Target territory A designated geographical area (or other parameter-based grouping) that is assigned to a sales group.

Tenancy Possession of a property by a tenant under the terms of a lease.

Tenant   A person who has temporary possession of a property under a lease or tenancy agreement.

Thought leader   A recognised authority in a specialised field.

Tone   How a person, group, organisation or issue is portrayed in the media; normally categorised as positive, neutral or negative, with various degrees of negative and positive tones.

Topping out   A ceremony traditionally held when the last beam (or its equivalent) is placed atop a structure during its construction; often used as a media event for PR purposes.

Unique selling point   The specific proposition that a brand or product offers to consumers.

Unique users   Unique IP addresses that have accessed a website. Calculating the number of unique users is a common way of measuring the popularity of a website.

Universal credit   A means-tested benefit for working-age people on a low income (UK).

Use Classes Order   The UK Town and Country Planning (Use Classes) Order 1987 puts uses of land and buildings into various categories. Planning permission is not needed for changes of use within the same use class unless a planning condition has been imposed restricting this.

User journey   A person's experience during a single website session, consisting of the series of actions performed to reach a specific page.

Values   Principles important to an organisation; often described alongside its mission and vision.

Video-on-demand   A video media distribution system that allows users to access video entertainment without a traditional video entertainment device and outside the constraints of a typical static broadcasting schedule.

Virtual reality   A simulated experience that can be similar to or completely different from the real world; can be used for entertainment or educational purposes.

Vision statement   A succinct, realistic, credible, easy-to-understand, relevant and ambitious description of the niche the organisation wants to occupy in the future.

Vlog   A blog created using video content, typically focused on a cause or special interest.

Web mapping   The process of using maps delivered by geographical information systems (GIS). A web map on the internet can allow interaction by the user.

Webinar   A seminar conducted online.

Wholly owned subsidiary (WOS)   A subsidiary company is considered wholly owned when another company (the parent company), owns all the common stock, there are no minority shareholders and the subsidiary's stock is not traded publicly. The term is commonly used in social housing, which typically operates with an 'umbrella' brand and a series of companies that specialise in providing either a specific product or for specific needs.

Wiki   A website combining the ongoing work of many authors, allowing users to modify the content of previous authors.

Wire services   A service for distributing a news release to the media and online. The main wire services are Business Wire, PR Newswire and PRWeb.

Yield   A measure of the return on an investment. A yield is the reciprocal of the multiplier that converts an income stream into a capital value.

Zero-carbon homes   Zero-carbon housing and zero-energy housing are terms used interchangeably to define single homes with a very high energy efficiency rating and low energy requirements.

Further reading

Andreasen, A. and Kotler, P. *Strategic Marketing for Nonprofit Organizations* (2008), New Jersey: Pearson Prentice Hall.

Beard, M. *Running a Public Relations Department* (2001), London: Kogan Page.

Black, C. *The PR Professional's Handbook: Powerful, Practical Communications* (2014), London: Kogan Page.

Black, S. *The Practice of Public Relations* (1995), Oxford: Butterworth-Heinemann.

Bland, M. *Communicating out of a Crisis* (1998), Basingstoke: Macmillan.

Bland, M., Theaker, A. and Wragg, D. *Effective Media Relations: How to Get Results* (2005), London: Kogan Page.

Brown, J., Gaudin, P. and Moran, W. *PR and Communication in Local Government and Public Services* (2013), London: Kogan Page.

Curtin, T. *Managing Green Issues* (2004), London: Palgrave Macmillan.

Green, A. *Creativity in Public Relations* (2009), London: Kogan Page.

Gregory, A. *Public Relations in Practice* (2003), London: Kogan Page.

Gregory, A. *Planning and Managing Public Relations Campaigns: A Strategic Approach* (2015), London: Kogan Page.

Griffin, A. *Crisis, Issues and Reputation Management* (2014), London: Kogan Page.

Grunig, J. *Excellence in PR and Communication Management* (1992). New Jersey: Lawrence Erlbaum.

Grunig, J. E. and Hunt, T. *Managing Public Relations* (1984), New York: Hunt, Holt, Rinehart & Winston.

Heath, R. L. *The Handbook of Public Relations* (2010), London: Sage.

Henslowe, P. *Public Relations: A Practical Guide to the Basics* (2003), London: Kogan Page.

Jefkins, F. (rev. Yadin, D.). *Public Relations* (1998), London: Pearson Professional.

Malpas, P. *Housing Associations and Housing Policy: A Historical Perspective* (2000), Basingstoke: Macmillan.

Morris, T. and Goldsworthy, S. *PR Today: The Authoritative Guide to Public Relations* (2016), London: Macmillan.

Norton, P. and Hughes, M. *Public Consultation and Community Involvement in Planning: a Twenty-First Century Guide* (2018), Oxford: Routledge.

Norwood, G. and Tasso, K. *Media Relations in Property* (2006), London: EG Books.

Oliver, S. M. *Public Relations Strategy* (2009), London: Kogan Page.

Parsons, P. J. *Ethics in Public Relations: A Guide to Best Practice* (2016), London: Kogan Page.

Reed, R. and Sims, S. *Property Development* (2014), Oxford: Routledge.

Rees, M. *Rethinking Social Housing* (2018), London: Chartered Institute of Housing.

Regester, M. and Larkin, J. *Risk Issues and Crisis Management in Public Relations: A Casebook of Best Practice* (2008), London: Kogan Page.

Signitzers B. and Olsen, T. *Using Communication Theory* (2000), London: Sage.

Thomson, S. and John, S. *Public Affairs in Practice: A Practical Guide to Lobbying* (2006), London: Kogan Page.

Waddington, S. *Chartered Public Relations Lessons from Expert Practitioners* (2015), London: Kogan Page.

Watson, T. and Noble, P. *Evaluating Public Relations: A Guide to Planning, Research and Measurement* (2014), London: Kogan Page.

Index

Printed in Great Britain
by Amazon

COMP
COOP

CSCW Cooperation or Conflict?
(ISBN 3-540-19755-9)
Steve Easterbrook (Ed.)

Computer Supported Collaborative Writing
(ISBN 3-540-19782-6)
Mike Sharples (Ed.)